W. Michaelis • Air Pollution

Springer
Berlin
Heidelberg
New York
Barcelona
Budapest
Hong Kong
London
Milan
Paris
Santa Clara
Singapore
Tokyo

W. Michaelis

Air Pollution

Dimensions, Trends, and Interactions with a Forest Ecosystem

With 89 Figures, 2 in Color and 26 Tables

Springer

PROFESSOR DR.RER.NAT.HABIL.WALFRIED MICHAELIS
GKSS Research Centre Geesthacht
Max-Planck-Straße
D-21502 Geesthacht
Germany

ISBN 3-540-61323-4 Springer-Verlag Berlin Heidelberg New York

Library of Congress Cataloging-in-Publication Data
Michaelis, W. (Walfried), 1931-
Air pollution : dimensions, trends, and interactions with a forest ecosystem / Walfried Michaelis.
p. cm.
Includes bibliographical references (p.) and index.
ISBN 3-540-61323-4 (hc)
1. Norway spruce – Effect of air pollution on–Germany–Hamburg Region. Norway spruce–Ecology–Germany–Hamburg Region. 3. Forest declines–Germany–Hamburg Region. 4. Air–Pollution–Environmental aspects–Germany–Hamburg Region. I. Title.
SB608.N67M53 1997
577. 3´276´0943512–dc21 96-49480

Cover Design: D & P GmbH, Heidelberg
Cover Illustration: W. Michaelis
Camera ready by UKT, Reichartshausen
SPIN 10534255 31/3137 5 4 3 2 1 0 - Printed on acid free paper -

Preface

In the early 1980s, forest decline became a matter of public and scientific concern when forest stands with Norway spruce (*Picea abies* [L.] Karst.) showed evident damage on a large geographical scale throughout Europe. The causes of the observed symptoms could not be elucidated on the basis of the state of knowledge at that time. Therefore, several research projects were launched both in Germany and in some other countries in order to identify the relevant pathogenic factors.

In 1985, the Federal Ministry for Research and Technology decided to include the site "Postturm", forest district Farchau/Ratzeburg, in the sponsorship of the research on forest decline as a site typical for lowlands and a sphere of anthropogenic urban influence. The investigation area is situated about 40 km east-northeast of the city of Hamburg. Since spruce trees in particular showed severe decline, emphasis was laid on this species. The programme started in 1986 and extended to 1992, with some activities continuing beyond this time. Working groups from 13 institutions took part in the overall project. The investigations concentrated on the following objectives:

- A comprehensive evaluation of air pollution including status, temporal trends and interaction with the forest ecosystem;
- the examination of possible direct relations between pollution and response of above-ground tree compartments;
- with respect to indirect impacts, a study of the influence of chemical soil quality on the element supply to the fine roots and other tree components;
- an assessment of the impact with regard to tree growth, wood formation and wood properties;
- the performance of fertilization experiments in order to explore the possibilities for stabilization of the forest stand, and finally
- with increasing progress in the field experiments a corroboration of the in-situ findings by studying model systems of spruce cultures under controlled laboratory conditions.

This volume concentrates on the first aspect which was the focal point of the research performed by the Institute of Physics of the GKSS Research Centre Geesthacht. However, other topics will also be treated, insofar as they are in close connection with the first subject and the GKSS working group was involved in these studies. An important feature of the project dealt with in this book was the aim not only to measure concentrations, but also fluxes with the final goal to derive flux balances. In this respect the project contrasts with other studies.

The author hopes that this volume will provide useful assistance in developing more profound insight into the complexity of air pollution and its interaction with a forest ecosystem.

Walfried Michaelis Geesthacht, January 1997

Acknowledgements

The research reported in this volume was sponsored by the Federal Ministry for Research and Technology and by the Environmental Ministry of the Schleswig-Holstein State Government. This support is gratefully acknowledged.

The author is very much indebted to the following colleagues for their dynamic and effective collaboration: R. Pepelnik, P. Rademacher, M. Schönburg, R.-P. Stößel and F. Theopold. Thanks are also due to H.-T. Mengelkamp for helpful discussions in connection with the meteorological aspects. The author also thanks R. Baumgart, M. Bormacher, W. Ehlers and H. Erbslöh for their active assistance in ion chromatography, trace element and gas analysis. Many thanks are also due to I. Eck for her effective support in producing the numerous figures of this volume.

The friendly cooperation with the colleagues from the other working groups of the overall project was a source of encouragement and inspiration. This is gratefully acknowledged. Particular thanks are due to J. Bauch for his work in coordinating the activities of the various research groups and for the many valuable discussions we had together during the course of this project.

Contents

1 Introduction

Since the mid-1970s damage to firs (*Abies alba* Mill.) in the form of needle chlorosis and needle loss has been observed in the Federal Republic of Germany. In the early 1980s forest stands with Norway spruce (*Picea abies* [L.] Karst.) showed the same symptoms on a large geographic scale. These findings gave rise to grave concern. Since the symptoms could not be explained on the basis of the knowledge at that time, both the Federal and the State governments initiated extensive research programmes to uncover the causes of the observed damage (Bayerische Staatsforstverwaltung 1982–1987; Ulrich 1983; Forschungsbeirat Waldschäden 1984, 1986; Projekt Europäisches Forschungszentrum für Maßnahmen zur Luftreinhaltung (PEF) 1985; Projektgruppe Bayern zur Erforschung der Wirkung von Umweltschadstoffen 1985a,b; Deutsche Forschungs- und Versuchsanstalt für Luft- und Raumfahrt (DFVLR) 1987; Stüttgen 1987; Minister für Umwelt, Raumordnung und Landwirtschaft des Landes NRW 1988 ff; Schulze et al. 1989). Slightly damaged trees may occur due to natural causes, for instance, due to climatic influences or epidemics, but more severe damage or even the death of trees was soon attributed to anthropogenic causes such as direct effects of air pollutants on plant organs, acid deposition, soil acidification and impairment of the mineral budget. Research priorities were created which focused on certain investigation sites and which led to manifold interdisciplinary cooperations. A similar development took place in other European countries where more and more forest decline was also observed (Bolhar-Nordenkampf 1989; Smidt et al. 1994; Landmann and Bonneau 1995).

In 1985, the German Federal Ministry of Research and Technology decided to include the project "Postturm", forest district Farchau/Ratzeburg, in the sponsorship as a site typical of lowlands and with an evident anthropogenic urban influence (Bauch and Michaelis 1988; Michaelis and Bauch 1992). This volume will deal with some of the central topics worked on during the Postturm project. The studies were performed by the Institute of Physics of the GKSS Research Centre Geesthacht.

The investigation site is situated in the North German State of Schleswig-Holstein and close to the city of Hamburg. From about 1981 in the region of Schleswig-Holstein damage to spruce had become evident, though at that time forest decline in South Germany was much more pronounced. However, at the Postturm site, severe damage to spruce trees was ascertained at this time. Hence, the diagnoses are a challenge for a comparison with the results obtained at various other sites extending from the State of Niedersachsen in North Germany down to the southern States of Bayern and Baden-Württemberg. Even before funding by the Federal Government began, research on various aspects such as tree growth, wood formation and wood properties was in progress. In 1985, an extended interdisciplinary research programme was elaborated and in 1986 the overall project finally started with 13 working groups. It officially ended in 1992, some activities extending beyond this date. A brief summary of the research concept and the coordination will be given in Chapter 3.

An indispensable precondition for the success of investigations on the problem of forest decline is an assessment of the impact of atmospheric pollutants. For this purpose it is desirable to consider a wide spectrum of different substances. Results of previous studies of the GKSS Research Centre showed excessive heavy metal concentrations in soil solutions from regions east and northeast of the city of Hamburg. Moreover, the close proximity of the Postturm investigation site to the large conurbation of Hamburg also suggested that various gaseous pollutants could be relevant to the research programme.

The focus of the present contribution is therefore a comprehensive evaluation of the air pollution including status, temporal variations, long-term trends and the interaction with the forest ecosystem. This task also necessarily required the examination of the direct relationship between pollution and the response of aboveground tree compartments as well as studies of the indirect impact via soil quality and element supply.

An important feature of the present study was not only to measure concentrations, but also the deposition and the fluxes between the various compartments with the final goal of deriving flux balances. These results alone ultimately allow conclusive statements with regard to the impact of atmospheric substances on the forest ecosystem. It is evident, however, that the transformation of the rather simple immission measurement technique to the analysis of fluxes requires considerable additional effort both in terms of equipment, methodology and data processing. Only subtle sampling methods as well as high precision and accuracy in the analytical techniques can ensure useful data and conclusive results. Numerous sensors are required for deriving meteorological auxiliary quantities. The availability of theoretical models and extensive computer programmes are further prerequisites. Furthermore, an interdisciplinary approach is indispensable. In the light of these features, the present project contrasts with other studies.

The construction of the measuring station was carried out in 1986. The Postturm tower itself was not well suited for the installation of the necessary equipment. Among others, two reasons were decisive for not using this tower: (1) its location was too close to the edge of the forest, and (2) due to postal installations the aerodynamic conditions above the canopy were adverse to deposition measurements. Therefore, a special tower was set up in order to ensure optimum preconditions. Gas measurements began in February 1987, and trace element investigations started in April 1987. Field experiments lasted until 1992, and the analysis of the comprehensive data sets took almost a further 2 years.

References

Bauch J, Michaelis W (eds)(1988) Das Forschungsprogramm Waldschäden am Standort "Postturm", Forstamt Farchau/Ratzeburg. GKSS Forschungszentrum Geesthacht, GKSS 88/E/55, 399 pp

Bayerische Staatsforstverwaltung (1982, 1983, 1984, 1985, 1986, 1987) Information 4/82, 4/83, 4/84, 4/85, 4/86, 4/87, München

Bolhar-Nordenkampf HR (ed) (1989) Streßphysiologische Ökosystemforschung Höhenprofil Zillertal. Phyton 29(3):1–302

Deutsche Forschungs- und Versuchsanstalt für Luft-und Raumfahrt (DFVLR) (ed) (1987) Statusseminar Untersuchung und Kartierung von Waldschäden mit Methoden der Fernerkundung. DFVLR-Druck Köln, 432 pp

Forschungsbeirat Waldschäden (1984) Zwischenbericht 1984. Literaturabteilung, Kernforschungszentrum Karlsruhe

Forschungsbeirat Waldschäden (1986) Bericht 1986. Literaturabteilung, Kernforschungszentrum Karlsruhe

Landmann G, Bonneau M (eds) (1995) Forest decline and atmospheric deposition effects in the French mountains. Springer, Berlin Heidelberg New York

Michaelis W, Bauch J (eds) (1992) Luftverunreinigungen und Waldschäden am Standort "Postturm", Forstamt Farchau/Ratzeburg. GKSS Forschungszentrum Geesthacht, GKSS 92/E/100, 455 pp

Minister für Umwelt, Raumordnung und Landwirtschaft des Landes NRW (ed) (1988 ff) Forschungsberichte zum Forschungsprogramm des Landes Nordrhein-Westfalen "Luftverunreinigungen und Waldschäden". No 1 ff, Düsseldorf

Projekt Europäisches Forschungszentrum für Maßnahmen zur Luftreinhaltung (PEF) (ed) (1985) 1. Statuskolloquium. Kernforschungszentrum Karlsruhe, KfK-PEF 2

Projektgruppe Bayern zur Erforschung der Wirkung von Umweltschadstoffen (PBWU) (1985a) Ergebnisse der Waldschadensforschung. Gesellschaft für Strahlen-und Umweltforschung, GSF-Bericht 8/85, Neuherberg

Projektgruppe Bayern zur Erforschung der Wirkung von Umweltschadstoffen (PBWU) (1985b) Atlas zur Waldschadensforschung. Gesellschaft für Strahlen-und Umweltforschung. GSF-Bericht 9/85, Neuherberg

Schulze ED, Lange OL, Oren R (eds) (1989) Forest decline and air pollution. Ecological Studies 77. Springer, Berlin Heidelberg New York

Smidt S, Herman F, Grill D, Guttenberger H (eds) (1994) Studies of ecosystems in the Limestone Alps – "Achenkirch Altitude Profiles". Phyton 34(3):1–192

Stüttgen E (ed) (1987) Statusseminar zum BMFT-Förderschwerpunkt Ursachenforschung zu Waldschäden. Kernforschungsanlage Jülich. Jül-Spez-413

Ulrich B (ed) (1983 ff) Berichte des Forschungszentrums Waldökosysteme/Waldsterben, Bd 1 ff. Universität Göttingen, Göttingen

2 The Investigation Site

The site "Postturm", forest district Farchau/Ratzeburg, is situated about 40 km east-northeast of the city of Hamburg (Fig. 2.1, site 2), and in the main wind direction from this conurbation. The landscape is characterized by an end moraine topography from the Weichsel (Würm) glacial period. The area shows a slight elevation with at most 20 m differences in altitudes, ranging between 60 and 80 m above sea level. The forest stand extends over an area of 1.4 km^2. On the mostly sandy and gravelly end moraine, slightly loamy sands with a medium silicate content make up the dominant component. The soil type which has developed is characterized by weakly to moderately podzolic braunerde. On this a 90- to 120-year-old mixed stand consisting predominantly of spruce (*Picea abies* [L.] Karst.) with some pine (*Pinus sylvestris* L.) and minor portions of beech (*Fagus sylvatica* L.) and oak (*Quercus robur* L.) exists (Bauch and Michaelis 1988; Michaelis and Bauch 1992).

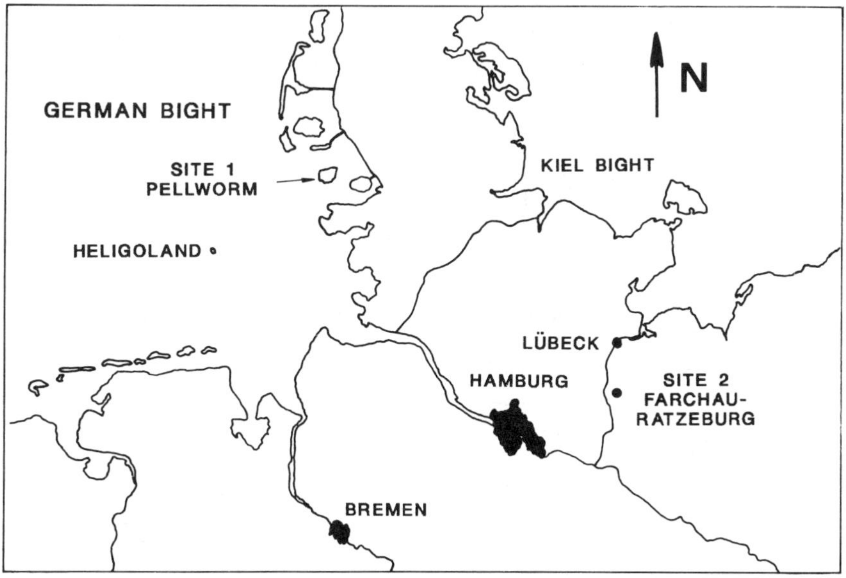

Fig. 2.1. Map of Schleswig-Holstein in North Germany with the investigation site "Postturm", Farchau/Ratzeburg

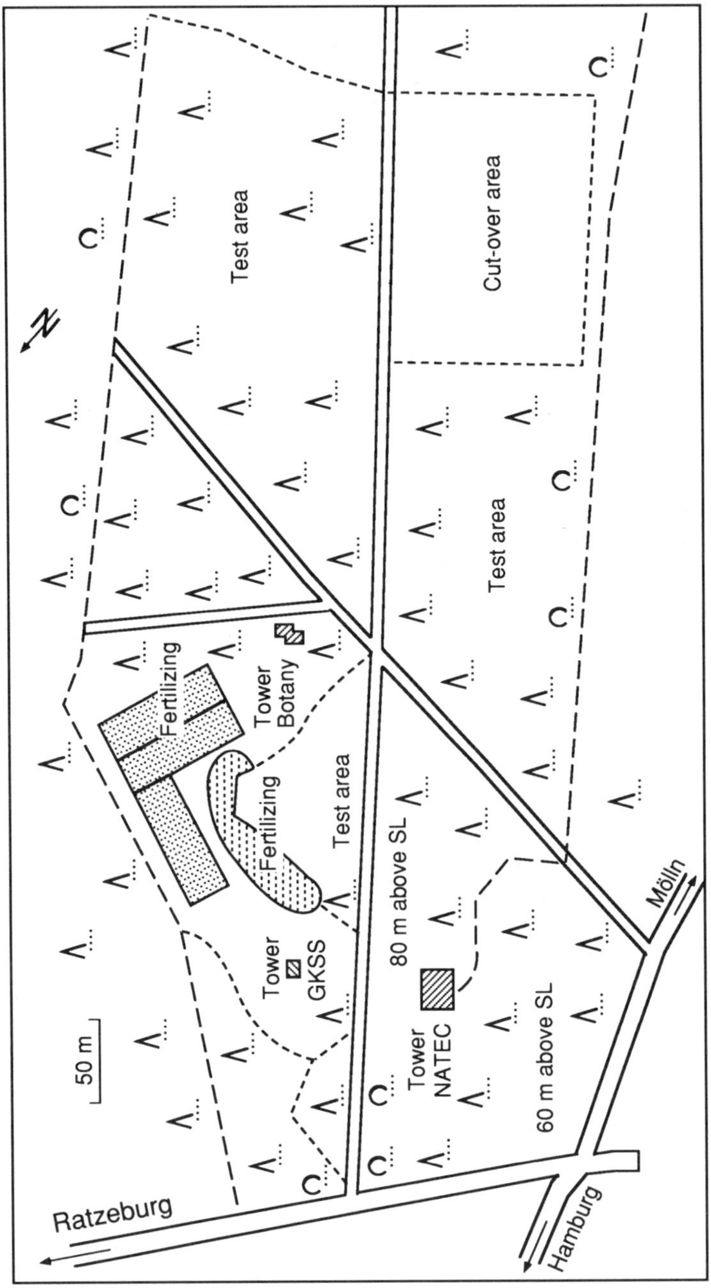

Fig. 2.2. Outline map of the investigation site with the three measuring stations and the test areas

Fig. 2.3. The radio tower of the Federal Postal Administration, the so-called Postturm, during the overall project also used as measuring station for organic pollutants

Due to the close proximity of Hamburg it was expected that the impact by atmospheric pollutants would be different from that observed at other investigation sites in Germany (Black Forest, Bavarian Forest, Fichtel Gebirge, Solling and Hils, respectively). In the early 1980s, the older spruce needles showed severe yellowing both at the top and bottom side. Since about 1954, a few trees have suffered an impairment of the wood accretion, a symptom which was markedly intensified following the very dry year of 1976. A similar reaction was exhibited by pine, but no effect was found during the 1980s in the case of beech and oak. Needle analyses, in part, very soon showed deficiencies in the elements calcium, potassium and magnesium.

Figure 2.2 presents an outline map of the investigation area. A radio tower of the Federal Postal Administration, the so-called Postturm (Fig. 2.3), lent its name to the overall project. This tower was used as equipment support for the detection of organic atmospheric pollutants (cf. Chap. 3). Since the Postturm station itself was not suited to carry out deposition and other flux measurements (cf. Chap. 1), a special tower was constructed for this purpose. The experiments focused mainly on recording concentrations and fluxes of inorganic constituents such as trace elements and gaseous pollutants and represent the main subject of this volume. A third station was used for biological investigations in the crown compartments of trees of different damage classes. Not indicated in Fig. 2.2 are numerous supplementary devices such as, for instance, the equipment used to measure throughfall, to take soil samples or to extract soil solutions. Finally, there were several proving grounds, in particular for diverse fertilizing experiments.

References

Bauch J, Michaelis W (eds) (1988) Das Forschungsprogramm Waldschäden am Standort "Postturm", Forstamt Farchau/Ratzeburg. GKSS Forschungszentrum Geesthacht, GKSS 88/E/55, 399 pp

Michaelis W, Bauch J (eds) (1992) Luftverunreinigungen und Waldschäden am Standort "Postturm" Forstamt Farchau/Ratzeburg. GKSS Forschungszentrum Geesthacht, GKSS 92/E/100, 455 pp

3 Research Concept and Coordination

The following institutions cooperated in the "Postturm" project:

1. Amt für Kreisforsten, D-23909 Farchau
2. Institut für Physik, GKSS-Forschungszentrum Geesthacht GmbH, Max-Planck-Straße, D-21502 Geesthacht
3. Institut für naturwissenschaftlich-technische Dienste GmbH (NATEC), Behringstraße 154, D-22763 Hamburg
4. Institut für Allgemeine Botanik und Botanischer Garten, Universität Hamburg, Ohnhorststraße 18, D-22609 Hamburg
5. Ordinariat für Holzbiologie, Universität Hamburg; Institut für Holzbiologie und Holzschutz, Bundesforschungsanstalt für Forst- und Holzwirtschaft, Leuschnerstraße 91, D-21031 Hamburg
6. Institut für Holzchemie und chemische Technologie des Holzes, Bundesforschungsanstalt für Forst- und Holzwirtschaft, Leuschnerstraße 91, D-21031 Hamburg
7. Ordinariat für Holztechnologie, Universität Hamburg; Institut für Holzphysik und mechanische Technologie des Holzes, Bundesforschungsanstalt für Forst- und Holzwirtschaft, Leuschnerstraße 91, D-21031 Hamburg
8. Institut für Forstgenetik und Forstpflanzenzüchtung, Bundesforschungsanstalt für Forst- und Holzwirtschaft, Sieker Landstraße 2, D-22927 Großhansdorf
9. Institut für Weltforstwirtschaft und Ökologie, Bundesforschungsanstalt für Forst- und Holzwirtschaft, Leuschnerstraße 91, D-21031 Hamburg
10. Institut für Biologische Informationsverarbeitung, Forschungszentrum Jülich GmbH, D-52425 Jülich
11. Forschungszentrum Waldökosysteme, Universität Göttingen, Büsgenweg 2, D-37077 Göttingen
12. Institut für Waldbau, Universität Göttingen, Büsgenweg 1, D-37077 Göttingen
13. Lehrstuhl für Spezielle Botanik und Botanischer Garten, Universität Tübingen, Auf der Morgenstelle 1, D-72076 Tübingen

The programme concentrated on five main themes (Bauch 1989). Since a comprehensive evaluation of air pollution is a basic prerequisite for the success of any study on forest decline, the first two of the measuring stations served for investigating inorganic and organic atmospheric substances, respectively. In the former case, both concentrations and fluxes were determined (Michaelis et al. 1988, 1989a, b, 1992a, b; Ellenberg and Kühnast 1988, 1992; Schönburg and Mengelkamp 1989; Michaelis and Theopold 1993; Pepelnik et al. 1993). A broad spectrum of organic trace constituents was measured at the other station (Dommröse and Figge 1988; Figge and Dommröse 1992).

In line with these studies, the second central point focused on the examination of the relationship between atmospheric pollution and alterations in the ecosystem, in particular, in the needles. The programme comprised the following aspects: quantification of the intake of organic pollutants into the needles and assessment of the cytotoxicity (field and laboratory experiments), detection of physiological stress indicators at the botanical measuring station, analyses of the trace element contents in needles and of fine-structural alterations in the needles of damaged trees (Figge 1988; Naumann et al. 1988; Meyberg and Kristen 1988; Meyberg et al. 1988a, b; Schulte-Baukloh et al. 1988; Kristen et al. 1992; Lalk et al. 1992a; Rademacher et al. 1992; Scheid and Dörffling 1992; Schmitt et al. 1992). During the realization of the project, with the aid of model systems, the studies were extended to investigations of secondary effects such as fungus disease and to exposure experiments with inorganic gaseous pollutants (Schell and Kristen 1992; Lalk et al. 1992b).

The third complex was devoted to the disclosure of the interrelations between chemical soil status, fine-root mass, properties and dynamics, and mycorrhiza functionality. The various contributions referred to the soil and fine-root inventory, to the influence of the element concentration in the soil solution on the element content in the roots and other compartments including the associated element fluxes and their balance, to the ion exchange in sterile and mycorrhiza fine roots and to their vitality. Different damage classes were considered and the in-situ experiments were supported by laboratory studies using model systems (Murach et al. 1988; Rademacher et al. 1988, 1992; Kuhn et al. 1988, 1992; Haug et al. 1992; Weber et al. 1992). Another aspect within this project complex focused on fertilizing experiments in various test areas with the aim of assessing their effect with regard to a stabilization of the forest stand (Dünisch et al. 1992; Rademacher and Kriebitzsch 1992).

The fourth central topic included investigations on the effect of air pollution with respect to tree growth, wood development and timber properties. The results of such studies are of great importance for the forestry and the wood industry with regard to the possibilities of utilization of wood from forest decline areas (Bauch et al. 1988; Eckstein 1988; Frühwald 1988; Frühwald and Schwab 1988; Göttsche-Kühn et al. 1988; Puls and Rademacher 1988; Dünisch and Bauch 1992; Krause and Eckstein 1992; Seehann 1992).

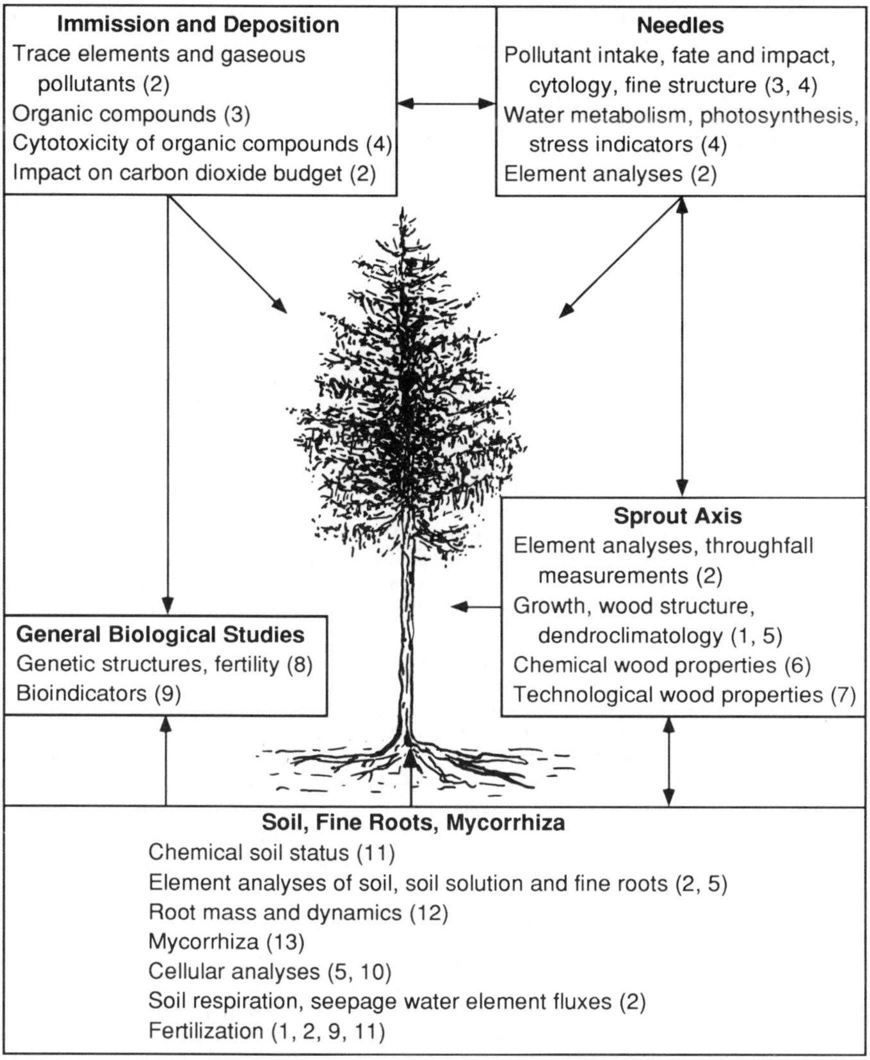

Fig. 3.1. Connections between the various projects. The *numbers in the brackets* refer to the list of institutions presented in the text

Finally, the comprehensive analysis of the stress factors at the investigation site offered the chance of examining possible alterations in the genetic structure of a forest stand (Geburek and Scholz 1988; Scholz 1988; Scholz and Venne 1988; Venne et al. 1988).

Figure 3.1 illustrates the cross-linkage between the individual research projects and also specifies the respective institutions listed at the beginning of this chapter. The present book gives a final account of the studies carried out by or in close cooperation with the Institute of Physics of the GKSS Research Centre Geesthacht. This refers particularly to the first complex outlined above, but various aspects of the second and the third subjects are also considered, since they directly interrelate with the atmospheric impact. Connections to other contributions of the "Postturm" project or to investigations performed elsewhere will be included in the discussion.

References

Bauch J (1989) Über das Forschungsprogramm Waldschäden am Standort "Postturm", Forstamt Farchau/Ratzeburg. In: Ulrich B (ed) International congress on forest decline research: state of knowledge and perspectives. Literaturabteilung, Forschungszentrum Karlsruhe, pp 567–582

Bauch J, Göttsche-Kühn H, Riehl G (1988) Zuwachsverlust bei Fichte auf der Waldschadensfläche "Postturm" im Forstamt Farchau/Ratzeburg. In: Bauch J, Michaelis W (eds) Das Forschungsprogramm Waldschäden am Standort "Postturm", Forstamt Farchau/Ratzeburg. GKSS Forschungszentrum Geesthacht, GKSS 88/E/55, pp 289–304

Dommröse AM, Figge K (1988) Qualitative und quantitative Bestimmung potentieller organischer Schadstoffe in der Luft des Standortes "Postturm", Forstamt Farchau/Ratzeburg. In: Bauch J, Michaelis W (eds) Das Forschungsprogramm Waldschäden am Standort "Postturm", Forstamt Farchau/Ratzeburg. GKSS Forschungszentrum Geesthacht, GKSS 88/E/55, pp 61–79

Dünisch O, Bauch J (1992) Fichtenkulturen als Modellsysteme für den Nachweis exogener Einflüsse auf das Baumwachstum. In: Michaelis W, Bauch J (eds) Luftverunreinigungen und Waldschäden am Standort "Postturm", Forstamt Farchau/Ratzeburg. GKSS Forschungszentrum Geesthacht, GKSS 92/E/100, pp 427–447

Dünisch O, Bauch J, Rademacher P, Puls J (1992) Beurteilung der Düngung eines umweltbelasteten Fichtenaltbestandes im Hinblick auf seine Stabilisierung. In: Michaelis W, Bauch J (eds) Luftverunreinigungen und Waldschäden am Standort "Postturm", Forstamt Farchau/Ratzeburg. GKSS Forschungszentrum Geesthacht, GKSS 92/E/100, pp 251–286

Eckstein D (1988) Dendroökologische Voruntersuchungen an Fichten in einem norddeutschen Waldschadensgebiet. In: Bauch J, Michaelis W (eds) Das Forschungsprogramm Waldschäden am Standort "Postturm", Forstamt Farchau/Ratzeburg. GKSS Forschungszentrum Geesthacht, GKSS 88/E/55, pp 305–310

Ellenberg H, Kühnast O (1988) Biomonitoring als Ansatz zur flächendeckenden Schadstoff-Erfassung: Schwermetallgehalte von Federproben der standorttreuen Vogelarten Elster (*Pica pica* L.) und Habicht (*Accipiter gentilis* L.) im Immissionsgradienten Hamburg-Ost. In: Bauch J, Michaelis W (eds) Das Forschungsprogramm Waldschäden am Standort "Postturm", Forstamt Farchau/Ratzeburg. GKSS Forschungszentrum Geesthacht, GKSS 88/E/55, pp 81–98

Ellenberg H, Kühnast O (1992) Zeitliche und räumliche Trends in den Depositionsraten von Schwermetallen im Raum östlich von Hamburg untersucht mit Hilfe von Habichtsfedern. In: Michaelis W, Bauch J (eds) Luftverunreinigungen und Waldschäden am Standort "Postturm", Forstamt Farchau/Ratzeburg. GKSS Forschungszentrum Geesthacht, GKSS 92/E/100, pp 61–70

Figge K (1988) Vorgang der ad-/absorptiven Anreicherung luftgetragener Organika in Fichtennadeln. In: Bauch J, Michaelis W (eds) Das Forschungsprogramm Waldschäden am Standort "Postturm", Forstamt Farchau/Ratzeburg. GKSS Forschungszentrum Geesthacht, GKSS 88/E/55, pp 101–119

Figge K, Dommröse AM (1992) Organische Spurenstoffe in der Atmosphäre der Waldstandorte "Postturm", Forstamt Farchau/Ratzeburg und "Donaustauf" bei Regensburg. In: Michaelis W, Bauch J (eds) Luftverunreinigungen und Waldschäden am Standort "Postturm", Forstamt Farchau/Ratzeburg. GKSS Forschungszentrum Geesthacht, GKSS 92/E/100, pp 71–90

Frühwald A (1988) Die Qualitätsentwicklung bei Trocken- und Naßlagerung von Nadelrundholz aus Waldschadensgebieten. In: Bauch J, Michaelis W (eds) Das Forschungsprogramm Waldschäden am Standort "Postturm", Forstamt Farchau/Ratzeburg. GKSS Forschungszentrum Geesthacht, GKSS 88/E/55, pp 349–363

Frühwald A, Schwab E (1988) Technologische Eigenschaften von Buchen und Eichen aus dem Waldschadensgebiet "Postturm" im Forstamt Farchau/Ratzeburg. In: Bauch J, Michaelis W (eds) Das Forschungsprogramm Waldschäden am Standort "Postturm", Forstamt Farchau/Ratzeburg. GKSS Forschungszentrum Geesthacht, GKSS 88/E/55, pp 337–348

Geburek T, Scholz F (1988) Versuch mit Waldbaumpopulationen von Fichte und Birke zur Viabilitätsselektion. In: Bauch J, Michaelis W (eds) Das Forschungsprogramm Waldschäden am Standort "Postturm", Forstamt Farchau/Ratzeburg. GKSS Forschungszentrum Geesthacht, GKSS 88/E/55, pp 383–387

Göttsche-Kühn H, Bauch J, Feuerstack M (1988) Anatomische Untersuchungen am Holz gesunder und geschädigter Fichten aus einem Waldschadensgebiet im Forstamt Farchau/Ratzeburg. In: Bauch J, Michaelis W (eds) Das Forschungsprogramm Waldschäden am Standort "Postturm", Forstamt Farchau/Ratzeburg. GKSS Forschungszentrum Geesthacht,. GKSS 88/E/55, pp 311–325

Haug I, Bauch H, Kottke I (1992) Die Mykorrhizierung von Fichten in Topfkulturen. In: Michaelis W, Bauch J (eds) Luftverunreinigungen und Waldschäden am Standort "Postturm", Forstamt Farchau/Ratzeburg. GKSS Forschungszentrum Geesthacht, GKSS 92/E/100, pp 409–425

Krause C, Eckstein D (1992) Holzzuwachs an Ästen, Stamm und Wurzeln bei normaler und extremer Witterung. In: Michaelis W, Bauch J (eds) Luftverunreinigungen und Waldschäden am Standort "Postturm", Forstamt Farchau/Ratzeburg. GKSS Forschungszentrum Geesthacht, GKSS 92/E/100, pp 215–242

Kristen U, Lockhausen J, Petersen W, Schult B, Strube K (1992) Veränderungen an Fichtennadeln nach Begasung mit 2,4-Dinitrophenol, Benzaldehyd, Furfural, Trichlorethan und Trichloressigsäure. In: Michaelis W, Bauch J (eds) Luftverunreinigungen und Waldschäden am Standort "Postturm", Forstamt Farchau/Ratzeburg. GKSS Forschungszentrum Geesthacht, GKSS 92/E/100, pp 341–352

Kuhn AJ, Schröder WH, Bauch J, Haug I, Kottke I, Oberwinkler F (1988) Mikrosonden-Untersuchungen zum Ionenaustausch und zur Ionenaufnahme in sterilen und mykorrhizierten Feinstwurzeln der Fichte. In: Bauch J, Michaelis W (eds) Das Forschungsprogramm Waldschäden am Standort "Postturm", Forstamt Farchau/Ratzeburg. GKSS Forschungszentrum Geesthacht, GKSS 88/E/55, pp 255–286

Kuhn AJ, Bauch J, Schröder WH (1992) Mikrosonden-Analysen zur Ionenaufnahme in Feinstwurzeln von Fichten (*Picea abies* [L.] Karst.) am Standort und in Modellsystemen. In: Michaelis W, Bauch J (eds) Luftverunreinigungen und Waldschäden am Standort "Postturm", Forstamt Farchau/Ratzeburg. GKSS Forschungszentrum Geesthacht,. GKSS 92/E/100, pp 365–408

Lalk I, Naumann R, Ludewig M, Fenner R, Dörffling K (1992a) Streßphysiologische Untersuchungen an Freilandfichten des Standortes "Postturm". In: Michaelis W, Bauch J (eds) Luftverunreinigungen und Waldschäden am Standort "Postturm", Forstamt Farchau/Ratzeburg. GKSS Forschungszentrum Geesthacht, GKSS 92/E/100, pp 93–118

Lalk I, Hartmann A, Dörffling K (1992b) Wirkung kurzzeitiger Schadgas-Expositionen (SO_2, NO_2, O_3) auf geklonte Jungfichten im Simulationsexperiment. In: Michaelis W, Bauch J (eds) Luftverunreinigungen und Waldschäden am Standort "Postturm", Forstamt Farchau/Ratzeburg. GKSS Forschungszentrum Geesthacht, GKSS 92/E/100, pp 309–340

Meyberg M, Kristen U (1988) Anwendung der Semidünnschnitt-Technik auf die histochemische Untersuchung von Nadelblättern. In: Bauch J, Michaelis W (eds) Das Forschungsprogramm Waldschäden am Standort "Postturm", Forstamt Farchau/Ratzeburg. GKSS Forschungszentrum Geesthacht, GKSS 88/E/55, pp 139–149

Meyberg M, Kappler R, Kristen U (1988a) Untersuchungen zur Cytotoxizität atmogener Organika des Standortes "Postturm". In: Bauch J, Michaelis W (eds) Das Forschungsprogramm Waldschäden am Standort "Postturm", Forstamt Farchau/Ratzeburg. GKSS Forschungszentrum Geesthacht, GKSS 88/E/55, pp 131–137

Meyberg M, Lockhausen J, Schult B, Kristen U (1988b) Ultrastrukturelle Veränderungen in Nadeln geschädigter Fichten des Standortes "Postturm". In: Bauch J, Michaelis W (eds) Das Forschungsprogramm Waldschäden am Standort "Postturm", Forstamt Farchau/Ratzeburg. GKSS Forschungszentrum Geesthacht, GKSS 88/E/55, pp 121–130

Michaelis W, Theopold F (1993) Deposition atmosphärischen Ozons und ihre Wirkung auf ein Waldökosystem. In: Arbeitsgemeinschaft der Großforschungseinrichtungen (AGF)(ed) Atmosphärisches Ozon – Prozesse und Wirkungen. Thenée Druck, Bonn, pp 25–27

Michaelis W, Schönburg M, Stößel RP (1988) Trocken- und Naßdeposition von Schwermetallen und Gasen. In: Bauch J, Michaelis W (eds) Das Forschungsprogramm Waldschäden am Standort "Postturm", Forstamt Farchau/Ratzeburg. GKSS Forschungszentrum Geesthacht, GKSS 88/E/55, pp 19–59

Michaelis W, Schönburg M, Stößel RP (1989a) Deposition of atmospheric pollutants into a North German forest ecosystem. In: Georgii HW (ed) Mechanisms and effects of pollutant-transfer into forests. Kluwer, Dordrecht, pp 3–12

Michaelis W, Schönburg M, Stößel RP (1989b) Schadstofftransfer in der Grenzschicht Atmosphäre-Vegetation. In: Arbeitsgemeinschaft der Großforschungseinrichtungen (AGF)(ed) Wechselwirkung Atmosphäre-Biosphäre. Thenée Druck, Bonn, pp 29–33

Michaelis W, Pepelnik R, Rademacher P, Riebesell M (1992a) Transfer of atmospheric pollutants into a forest ecosystem. In: Teller A, Mathy P, Jeffers JNR (eds) Responses of forest ecosystems to environmental changes. Elsevier, London, New York, pp 596–597

Michaelis W, Pepelnik R, Theopold F, Rademacher P (1992b) Deposition atmosphärischer Spurenstoffe und Stoffflüsse im Ökosystem Wald. In: Michaelis W, Bauch J (eds) Luftverunreinigungen und Waldschäden am Standort "Postturm", Forstamt Farchau/Ratzeburg. GKSS Forschungszentrum Geesthacht, GKSS 92/E/100, pp 11–59

Murach D, Rapp C, Ulrich B (1988) Boden- und Feinwurzelinventur auf den Versuchsflächen am Standort "Postturm", Forstamt Farchau/Ratzeburg. In: Bauch J, Michaelis W (eds) Das Forschungsprogramm Waldschäden am Standort "Postturm", Forstamt Farchau/Ratzeburg. GKSS Forschungszentrum Geesthacht, GKSS 88/E/55, pp 189–214

Naumann R, Ludewig M, Fenner R, Lalk I, Bigdon M, Dörffling K (1988) Untersuchung physiologischer Kenngrößen zur Beurteilung des Schadenszustandes von Fichten. In: Bauch J, Michaelis W (eds) Das Forschungsprogramm Waldschäden am Standort "Postturm", Forstamt Farchau/Ratzeburg. GKSS Forschungszentrum Geesthacht, GKSS 88/E/55, pp 151–171

Pepelnik R, Erbslöh B, Michaelis W, Prange A (1993) Determination of trace element deposition into a forest ecosystem using total-reflection X-ray fluorescence. Spectrochim Acta 48B(2):223–229

Puls J, Rademacher P (1988) Jahreszeitliche Schwankungen im Gehalt von löslichen Zuckern, Stärke und Extraktstoffen im Holz gesunder und geschädigter Fichten (*Picea abies* [L.] Karst.). In: Bauch J, Michaelis W (eds) Das Forschungsprogramm Waldschäden am Standort "Postturm", Forstamt Farchau/Ratzeburg. GKSS Forschungszentrum Geesthacht, GKSS 88/E/55, pp 327–336

Rademacher P, Kriebitzsch WU (1992) Diagnostischer Düngungsversuch an Fichte am Standort "Postturm". In: Michaelis W, Bauch J (eds) Luftverunreinigungen und Waldschäden am Standort "Postturm", Forstamt Farchau/Ratzeburg. GKSS Forschungszentrum Geesthacht, GKSS 92/E/100, pp 287–306

Rademacher P, Bauch J, Michaelis W (1988) Einfluß der Elementkonzentration der Bodenlösung auf den Elementgehalt in gesunden und geschädigten Fichten des Standortes "Postturm". In: Bauch J, Michaelis W (eds) Das Forschungsprogramm Waldschäden am Standort "Postturm", Forstamt Farchau/Ratzeburg. GKSS Forschungszentrum Geesthacht, GKSS 88/E/55, pp 215–254

Rademacher P, Ulrich B, Michaelis W (1992) Bilanzierung der Elementvorräte und Elementflüsse innerhalb der Ökosystemkompartimente Krone, Stamm, Wurzel und Boden eines belasteten Fichtenbestandes am Standort "Postturm". In: Michaelis W, Bauch J (eds) Luftverunreinigungen und Waldschäden am Standort "Postturm", Forstamt Farchau/Ratzeburg. GKSS Forschungszentrum Geesthacht, GKSS 92/E/100, pp 149–186

Scheid P, Dörffling K (1992) Jahreszeitliche Veränderungen im Gehalt des Phytohormons Indol-3-essigsäure in der Kambialregion von Fichten am Standort "Postturm", Forstamt

Farchau/Ratzeburg, in Beziehung zur Kronenverlichtung und zur Jahrringbreite. In: Michaelis W, Bauch J (eds) Luftverunreinigungen und Waldschäden am Standort "Postturm", Forstamt Farchau/Ratzeburg. GKSS Forschungszentrum Geesthacht, GKSS 92/E/100, pp 119–137

Schell R, Kristen U (1992) Trichloressigsäure begünstigt Pilzinfektionen von Fichtennadeln. In: Michaelis W, Bauch J (eds) Luftverunreinigungen und Waldschäden am Standort "Postturm", Forstamt Farchau/Ratzeburg. GKSS Forschungszentrum Geesthacht, GKSS 92/E/100, pp 353–363

Schmitt U, Kristen U, Ruetze M, Schultze R (1992) Zur Feinstruktur von Mesophyll- und Siebzellen in Nadeln gering und stark geschädigter Fichten (*Picea abies* [L.] Karst.) des Standortes "Postturm"/Ratzeburg. In: Michaelis W, Bauch J (eds) Luftverunreinigungen und Waldschäden am Standort "Postturm", Forstamt Farchau/Ratzeburg. GKSS Forschungszentrum Geesthacht, GKSS 92/E/100, pp 139–148

Scholz F (1988) Zur Bedeutung populationsgenetischer Untersuchungen in der Waldschadensforschung. In: Bauch J, Michaelis W (eds) Das Forschungsprogramm Waldschäden am Standort "Postturm", Forstamt Farchau/Ratzeburg. GKSS Forschungszentrum Geesthacht, GKSS 88/E/55, pp 367–369

Scholz F, Venne H (1988) Untersuchung von Genotyp-Umwelt-Interaktionen mit Fichtenklonen auf unterschiedlichen Standorten sowie unter Begasung. In: Bauch J, Michaelis W (eds) Das Forschungsprogramm Waldschäden am Standort "Postturm", Forstamt Farchau/Ratzeburg. GKSS Forschungszentrum Geesthacht, GKSS 88/E/55, pp 371–382

Schönburg M, Mengelkamp HT (1989) Turbulente Flüsse in der atmosphärischen Grenzschicht über einem Waldgebiet. In: Arbeitsgemeinschaft der Großforschungseinrichtungen (AGF) (ed) Wechselwirkung Atmosphäre-Biosphäre. Thenée Druck, Bonn, pp 5–7

Schulte-Baukloh C, Fenner R, Ludewig M, Naumann R, Lalk I, Bigdon M, Dörffling K (1988) Untersuchungen zur Tagesperiodik der interzellulären Ethylenkonzentration an Fichtentrieben. In: Bauch J, Michaelis W (eds) Das Forschungsprogramm Waldschäden am Standort "Postturm", Forstamt Farchau/Ratzeburg. GKSS Forschungszentrum Geesthacht, GKSS 88/E/55, pp 173–185

Seehann G (1992) Stammfäule an Fichten unterschiedlicher Schädigung eines Waldschadensstandortes im Forstamt Farchau/Ratzeburg. In: Michaelis W, Bauch J (eds) Luftverunreinigungen und Waldschäden am Standort "Postturm", Forstamt Farchau/Ratzeburg. GKSS Forschungszentrum Geesthacht, GKSS 92/E/100, pp 243–247

Venne H, Scholz F, Bergmann F, Geburek T (1988) Populationsgenetische Untersuchungen an Altbäumen, deren Samenkollektiven und Nachkommen auf immissionsbelasteten Standorten. In: Bauch J, Michaelis W (eds) Das Forschungsprogramm Waldschäden am Standort "Postturm", Forstamt Farchau/Ratzeburg. GKSS Forschungszentrum Geesthacht, GKSS 88/E/55, pp 389–395

Weber R, Kottke I, Oberwinkler F (1992) Fluoreszenzmikroskopische Untersuchungen zur Vitalität von Mykorrhizen an Fichten (*Picea abies* [L.] Karst.) verschiedener Schadklassen am Standort "Postturm", Forstamt Farchau/Ratzeburg. In: Michaelis W, Bauch J (eds) Luftverunreinigungen und Waldschäden am Standort "Postturm", Forstamt Farchau/-Ratzeburg. GKSS Forschungszentrum Geesthacht, GKSS 92/E/100, pp 187–211

4 General Outline of the Measuring Station

From the outset, the present study was designed to determine fluxes (cf. Chap. 1). A special tower was erected for the installation of the equipment which is necessary to take samples of the various pollutants and to derive the appropriate meteorological auxiliary quantities (Michaelis et al. 1988, 1989a,b, 1990, 1992b,c; Michaelis and Theopold 1993). The tower was a 48-m-high triangular lattice mast with favourable aerodynamic properties. It stood on a concrete base and was braced at four altitudes to three gravel-filled 10 m^3 containers by 10 mm diameter steel cables. It was positioned 124 m northeast of the "Postturm" in an area characterized by a passably preserved tree population, so that the conditions for deposition measurements were rather favourable, in contrast to the conditions at the postal tower where the stand was already markedly cleared and where representative deposition measurements would have been practically impossible. The treetop height at the measuring station was about 26 m.

For the determination of the deposition of heavy metals and other relevant trace elements, automatic rainwater and aerosol samplers were used. These instruments can be employed even at temperatures as low as −30 °C. The dry deposition was derived by means of the so-called concentration method (cf. Chap. 6). This required furnishing the aerosol sampler with a cascade impactor for particle-size fractionating. The flux to the surface then follows from the sum of the particle size-dependent products of concentration and deposition velocity. The latter quantity must be deduced either from theoretical deposition models or from results of special experimental studies (Sehmel and Hodgson 1978; Jonas and Vogt 1982; Höfken and Gravenhorst 1983; Jonas 1984; Brückmann 1988; Grosch and Schmitt 1988; Waraghai and Gravenhorst 1989; Hertlein 1990).

Advanced analytical methods were applied in order to determine a broad spectrum of atmospheric constituents in both rainwater and aerosols. Highly efficient multielement detection was performed using total-reflection X-ray fluorescence and inductively coupled plasma optical emission spectroscopy (Stößel and Prange 1985; Michaelis 1986b; Michaelis and Prange 1988; Michaelis et al. 1992a; Prange and Schwenke 1992; Pepelnik et al. 1993). Most of the ions in rainwater were determined by means of ion chromatography.

The gas measurements included the gases NO, NO_2, NH_3, SO_2, O_3 and CO_2. For determining the dry deposition, the so-called gradient method was applied (cf. Chap. 6). Therefore the tower was equipped with five heated, outside insulated and inside Teflon-coated suction pipes which took air samples at approximately 47, 36, 28, 9 and 1 m height. The pipe at 1 m was installed at a later phase of the project in order to obtain additional information on the fluxes at the air-soil interface. Gas analysis was performed using chemiluminescence, UV-excited fluorescence, UV photometry and infrared absorption.

The evaluation and interpretation of the analytical data by means of the concentration and gradient method require the knowledge of numerous meteorological parameters. For their determination the following instruments were mounted on the tower: eight anemometers (measuring range to 41 m/s), five electrically ventilated thermometers after Frankenberger (measuring range −35 to +45 °C), five hair hygrometers (measuring range 5 to 100%), one linear wind direction sensor, one sternpyranometer after Dirmhirn for measuring the global radiation (range 0 to 1330 W/m^2), one precipitation gauge and one barometer (range 950 to 1050 hPa), altogether 22 sensors. The anemometers were mounted on cantilevers, two at each of the upper four heights in a north-south direction, in order to minimize impairment of the data by the tower structure. Critical sensors were sheltered from freezing by automatic heating.

The main features of the techniques applied in the present study had already stood the test in the course of a former project which had been performed on the Island of Pellworm (Fig. 2.1, site 1) with the goal of quantifying the wet and dry deposition of pollutants into the German Bight. These data then served as a basis for estimating the atmospheric contribution to the pollution of the North Sea (Michaelis 1986a, 1987; Michaelis and Stößel 1986; Stößel 1987). During the present study, the techniques were continuously further improved.

A star-shaped gutter system below the crown area was used for throughfall measurements. Samples of soil solutions were taken by means of the lysimeter technique. A schematic view of the measuring station is shown in Fig. 4.1, demonstrating in particular the instrumentation for sampling and meteorological measurements. A photograph of the tower is presented in Fig. 4.2. Two containers provided the infrastructure of the measuring station. In the first one, the rainwater and aerosol samples were prepared for the transport to the trace analytical laboratory at Geesthacht. The other one contained the gas analyzers, electronics and the central computer for controlling the continuous experiments (Fig. 4.3).

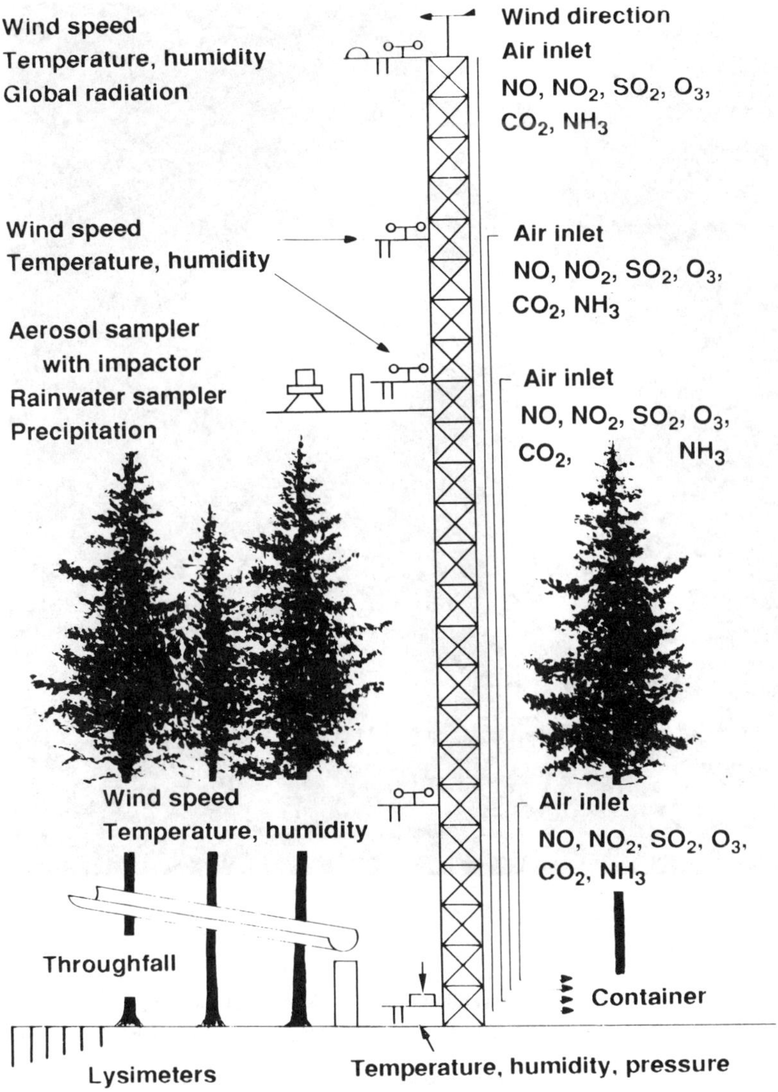

Fig. 4.1. Schematic view of the measuring station

References

Brückmann A (1988) Radionuklidbilanz von 4 Waldökosystemen nach dem Reaktorunfall in Tschernobyl und eine Bestimmung der trockenen Deposition. Diplomarbeit, Forstwissenschaftlicher Fachbereich, Universität Göttingen

Grosch S, Schmitt G (1988) Experimental investigations on the deposition of trace elements in forest areas. In: Grefen K, Löbel J (eds) Environmental meteorology. Kluwer, Dordrecht, pp 201–216

Fig. 4.2. Lattice mast with instrumentation for sampling and meteorological measurements

Hertlein F (1990) Untersuchungen zur Anwendung der Gradientenmethode auf luftgetragene Partikel. Diplomarbeit, Fachbereich Angewandte Naturwissenschaften, Fachhochschule Lübeck

Höfken KD, Gravenhorst G (1983) Untersuchung über die Deposition atmosphärischer Spurenstoffe in Buchen- und Fichtenwald. In: UBA-Berichte 6/83, Teil II. Schmidt-Verlag, Berlin

Jonas R (1984) Ablagerung und Bindung von Luftverunreinigungen an Vegetation und anderen atmosphärischen Grenzflächen. Kernforschungsanlage Jülich, KFA-Jül-1949

Jonas R, Vogt KJ (1982) Untersuchungen zur Ermittlung der Ablagerungsgeschwindigkeit von Aerosolen auf Vegetation und anderen Probenahmeflächen. Kernforschungsanlage Jülich. KFA-Jül-1780

Michaelis W (1986a) Naß- und Trockendeposition von Schwermetallen. In: Bodenschutz – Lösung durch Technik. Tech Mitt 79, Haus der Technik e.V., Essen, pp 266–271

Michaelis W (1986b) Multielement analysis of environmental samples by total-reflection X-ray fluorescence spectrometry, neutron activation analysis and inductively coupled plasma optical emission spectroscopy. Fresenius Z Anal Chem 324:662–671

Fig. 4.3. Facilities for deposition studies. *Centre* Base of the lattice mast with gas suction pipes. *Left* Container for sample preconditioning; *right* container with gas analyzers, electronics and process computer

Michaelis W (1987) Experimental studies on dry deposition of heavy metals and gases. 16th NATO/CCMS Int Tech Meet on Air pollution modeling and its application, April 6–10, 1987, Lindau. Also in: Van Dop H (ed) Air pollution modeling and its application VI. Plenum Press, New York, 1988, pp 61–74

Michaelis W, Prange A (1988) Trace analysis of geological and environmental samples by total-reflection X-ray fluorescence spectrometry. Nucl Geophys 2(4):231–245

Michaelis W, Stößel RP (1986) Untersuchungen zur Schwermetalldeposition auf der Insel Pellworm – ein Beitrag zur Ermittlung des atmosphärischen Schadstoffeintrags in die Nordsee. In: GKSS Annu Rep 1986, GKSS Forschungszentrum Geesthacht, pp 8–23

Michaelis W, Theopold F (1993) Deposition atmosphärischen Ozons und ihre Wirkung auf ein Waldökosystem. In: Arbeitsgemeinschaft der Großforschungseinrichtungen (AGF)(ed) Atmosphärisches Ozon – Prozesse und Wirkungen. Thenée Druck, Bonn, pp 25–27

Michaelis W, Schönburg M, Stößel RP (1988) Trocken- und Naßdeposition von Schwermetallen und Gasen. In: Bauch J, Michaelis W (eds) Das Forschungsprogramm Waldschäden am Standort "Postturm", Forstamt Farchau/Ratzeburg. GKSS Forschungszentrum Geesthacht, GKSS 88/E/55, pp 19–59

Michaelis W, Schönburg M, Stößel RP (1989a) Deposition of atmospheric pollutants into a North German forest ecosystem. In: Georgii HW (ed) Mechanisms and effects of pollutant-transfer into forests. Kluwer, Dordrecht, pp 3–12

Michaelis W, Schönburg M, Stößel RP (1989b) Schadstofftransfer in der Grenzschicht Atmosphäre-Vegetation. In: Arbeitsgemeinschaft der Großforschungseinrichtungen (AGF) (ed) Wechselwirkung Atmosphäre-Biosphäre, Thenée Druck, Bonn, pp 29–33

Michaelis W, Pepelnik R, Rademacher P, Riebesell M (1990) Wechselwirkung zwischen Luftschadstoffen und Vegetation. In: GKSS Annu Rep 1990, GKSS Forschungszentrum Geesthacht, pp 42–55

Michaelis W, Pepelnik R, Prange A (1992a) Application of TXRF in environmental research. PICXAM – Pacific International Congress on X-Ray Analytical Methods, August 7–16,

1991, Hilo and Honolulu. In: Barret CS, Gilfrich JV, Huang TC, Jenkins R, McCarthy GJ, Predecki PK, Ryon R, Smith DK (eds) Advances in X-ray analysis, vol 35B. Plenum Press, New York, London, pp 953–958

Michaelis W, Pepelnik R, Rademacher P, Riebesell M (1992b) Transfer of atmospheric pollutants into a forest ecosystem. In: Teller A, Mathy P, Jeffers JNR (eds) Responses of forest ecosystems to environmental changes. Elsevier, London, New York, pp 596–597

Michaelis W, Pepelnik R, Theopold F, Rademacher P (1992c) Deposition atmosphärischer Spurenstoffe und Stoffflüsse im Ökosystem Wald. In: Michaelis W, Bauch J (eds) Luftverunreinigungen und Waldschäden am Standort "Postturm", Forstamt Farchau/Ratzeburg. GKSS Forschungszentrum Geesthacht, GKSS 92/E/100, pp 11–59

Pepelnik R, Erbslöh B, Michaelis W, Prange A (1993) Determination of trace element deposition into a forest ecosystem using total-reflection X-ray fluorescence. Spectrochim Acta 48B(2):223–229

Prange A, Schwenke H (1992) Trace element analysis using total-reflection X-ray fluorescence spectrometry. PICXAM – Pacific International Congress on X-Ray Analytical Methods, August 7–16, 1991, Hilo and Honolulu. In: Barret CS, Gilfrich JV, Huang TC, Jenkins R, McCarthy GJ, Predecki PK, Ryon R, Smith DK (eds) Advances in X-ray analysis, vol 35B. Plenum Press, New York, pp 899–923

Sehmel GA, Hodgson WH (1978) A model for predicting dry deposition of particles and gases to environmental surfaces. Batelle Richland, Washington, PNL-SA-6721

Stößel RP (1987) Untersuchungen zur Naß-und Trockendeposition von Schwermetallen auf der Insel Pellworm. GKSS Forschungszentrum Geesthacht, GKSS 87/E/34

Stößel RP, Prange A (1985) Determination of trace elements in rainwater by total-reflection X-ray fluorescence. Anal Chem 57:2880–2885

Waraghai A, Gravenhorst G (1989) Dry deposition of atmospheric particles to an old spruce stand. In: Georgii HW (ed) Mechanismus and effects of pollutant-transfer into forests. Kluwer, Dordrecht, pp 77–86

5 Experimental Procedures

5.1 Sampling Techniques

5.1.1 Rainwater Sampler

In addition to direct interaction, atmospheric pollutants are deposited on the surface of the earth via precipitation by the processes of washout and rainout. From a scientific point of view, in particular with regard to the objective of measuring fluxes into and within the ecosystem, including their balances, each component has to be quantified, i.e. it is indispensable to separate as far as possible wet and dry deposition. One of the devices required is the so-called wet-only sampler. The demands can be quite well fulfilled by automatic devices which by means of a moisture sensor are opened and closed at the start and at the end of rainfall, respectively. Strictly speaking, the 'wet deposition' has to be defined as the amount of a constituent which is collected during the opening of the sampler. As a consequence, in particular during events with very low precipitation intensity, the results may include non-negligible contributions of the dry deposition. Therefore, the moisture sensor has to be very efficient.

The wet-only-sampler used in the present study (Bleymehl-Reinraumtechnik, Jülich) originates from a development in the Research Centre Jülich (Nguyen and Valenta 1978). A sketch of the device is given in Fig. 5.1. The active receiver area is 598 cm^2 and thus well above the recommended threshold value. Heating of the receiver and the lid ensures reliable operation down to temperatures of −30 °C. An electronic control system activates a motor for opening and closing the sampler. The corresponding orders are given by the precipitation sensor. During rainfall, this sensor is heated in order to achieve efficient operation. Investigations on the drop-size distribution of precipitation have revealed a threshold intensity of about 0.03 to 0.05 mm/h at which a sensor should respond during the first 2 min (Winkler 1993a; Winkler et al. 1993). Within the scope of a thorough intercomparison of various sensor types, it was shown that the Bleymehl sensor with 0.05 mm/h under outdoor conditions complies with this demand (Winkler 1993b). Nevertheless, during rainfall events, small droplets may escape sampling, an effect which depends on the wind velocity and which, due to

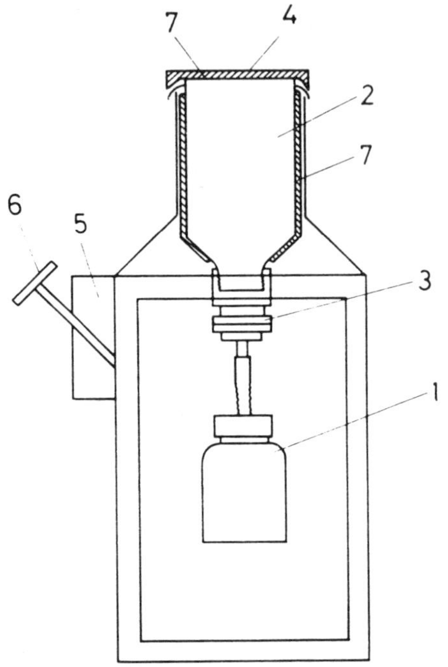

Fig. 5.1. Wet-only sampler. *1* Collecting reservoir; *2* receiver; *3* filter assembly; *4* lid; *5* electric motor; *6* precipitation sensor; *7* heating

the wind profile, is of particular importance in the case of measurements above a forest canopy. In order to reduce possible errors caused by this effect, a correction procedure was applied on the basis of parallel measurements with one wet-only sampler on the tower and a second one properly positioned in a nearby open terrain. Losses by wetting and subsequent drying in the receiver (Winkler et al. 1989) were avoided to a large extent by determining the amount of rainfall with a separate precipitation gauge and, concerning the rainwater trace constituents, by subjecting the receiver to a careful rinsing process at the end of each sampling period.

Undissolved constituents were strained off by means of a 0.45 μm membrane filter and were analysed separately. The collecting reservoir was completely protected against light irradiation, in order to minimize alterations in the sample. The size of the reservoir could be adjusted to between 1 and 10 l depending on the length of the sampling period. Apart from special parallel investigations, weekly samples were taken throughout the whole project. All relevant components of the device were made of polyethylene. For control purposes a permanently open sampler was also operated on the tower with an identical time sequence (Michaelis et al. 1988, 1989, 1992a,c).

5.1.2 Throughfall Measurements

Due to the inhomogeneous structure of the crown compartment, representative sampling of throughfall constitutes a particular problem. Small-scale variations can seriously falsify the results. This is valid for both the amount and the chemical composition of the throughfall. One possibility of solving this problem is the employment of a great number of samplers properly positioned within the forest. However, this procedure involves a considerable financial and working expenditure which can only be accomplished, inter alia, in the case of long sampling times and restricted trace analyses.

In order to avoid these restrictions, at a carefully selected position a sampling system was set up which consisted of three 4-m-long gutters in a star-shaped arrangement and optionally two collecting funnel units (Fig. 5.2). The gutters were made of polyvinyl chloride and coated with a self-adhesive Teflon foil. Each of them had a sampling area of 0.5 m^2 so that together an area of 1.5 m^2 represented an overall plane of about 50 m^2. The inclination angle of the gutters was 10°. All relevant components of the collecting units were made of polyethylene. As in the case of the wet-only sampler on the tower (cf. Sect. 5.1.1), undissolved constituents were strained off by means of 0.45-μm membrane filters and analysed separately. Further

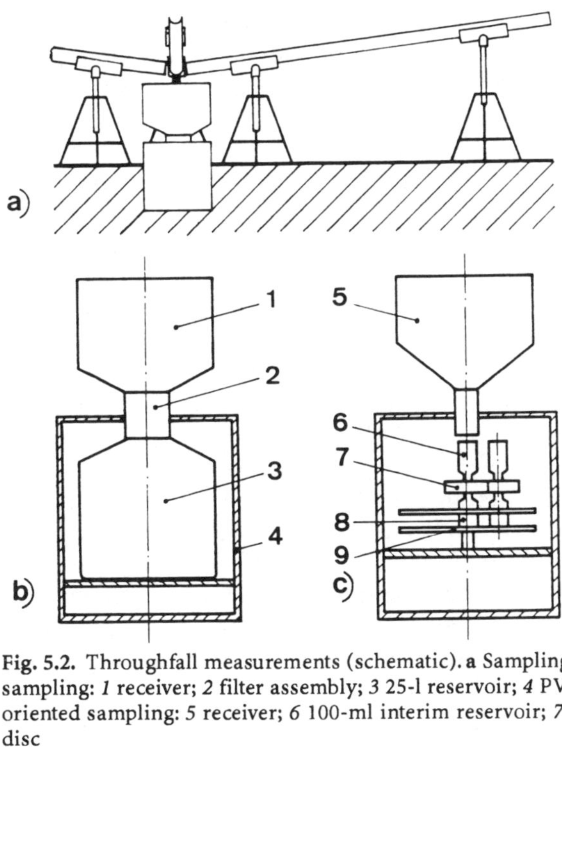

Fig. 5.2. Throughfall measurements (schematic). **a** Sampling system. **b** Accumulator for weekly sampling: *1* receiver; *2* filter assembly; *3* 25-l reservoir; *4* PVC casing. **c** Accumulator for event-oriented sampling: *5* receiver; *6* 100-ml interim reservoir; *7* filter; *8* 100-ml reservoir; *9* rotary disc

details, in particular those regarding the cleaning procedures, may be found elsewhere (Panten 1990; Schultz 1991).

5.1.3 Sampling of Size-Fractionated Particulates

To determine the dry deposition of airborne particulates several methods in principle may be considered (Winkler 1985). Expenditure and the chance of success differ, however, markedly and systematic experimental intercomparisons of the precision and accuracy are scarce. Basically, the quantification of particle fluxes is much more intricate than that of wet deposition. This is due to the complex interaction mechanisms involved which include sedimentation, turbulent diffusion as well as Brownian motion.

A promising procedure is provided by the so-called concentration method (Michaelis 1986a; Stößel 1986). In this case, the flux to the surface follows from the sum over the products of the particle size-dependent quantities concentration and deposition velocity. Details of the method and the problem of determining the deposition velocity as a function of the aerodynamic particle diameter will be discussed in Chapter 6. In addition, another important task is the sampling of size-fractionated particulates.

For this purpose a high-volume sampler (HVS-150, Ströhlein Instruments, Kaarst) equipped with a five-stage slotted cascade impactor (Model 235, Sierra Instruments, Inc., Carmel Valley) was employed during the present study (Michaelis et al. 1992a). The flow rate was 69 m^3/h. Cellulose paper filters were used as collection substrates. Six fractions of aerosols, including the back-up filter, were obtained in this way.

The main problem with this technique is the question of how real airborne particulates behave with respect to the collection efficiencies of the various impactor stages. Investigations of the sampling characteristics have shown that droplets and solid particles reveal quite different results (Röbig et al. 1980). There is strong evidence that real environmental aerosols exhibit a behaviour very similar to that of droplets, since the aerosols have a rather good adhesiveness due to the formation of a hydrate coating in the humid outdoor air (Winkler 1974; Rao and Whitby 1978; Dannecker et al. 1982). The collection efficiencies for each stage which underlie the present study are summarized in Fig. 5.3. Specifications given by the manufacturer and experimental data (Röbig et al. 1980) are presented together with mathematical approximation functions which were used for the evaluation. The calibration was obtained with the aid of monodisperse oleic acid particles. The numbers 1 to 5 refer to the individual impactor stages. Curve 0 describes the presegregation of coarse particles caused by the upstream inlet of the sampler. By this process raised dust of mineral origin is prevented from being collected. All curves are plotted as a function of the equivalent aerodynamic diameter in order to facilitate their use in applications. Taking the presegregation into account, from the products of the dotted curves in

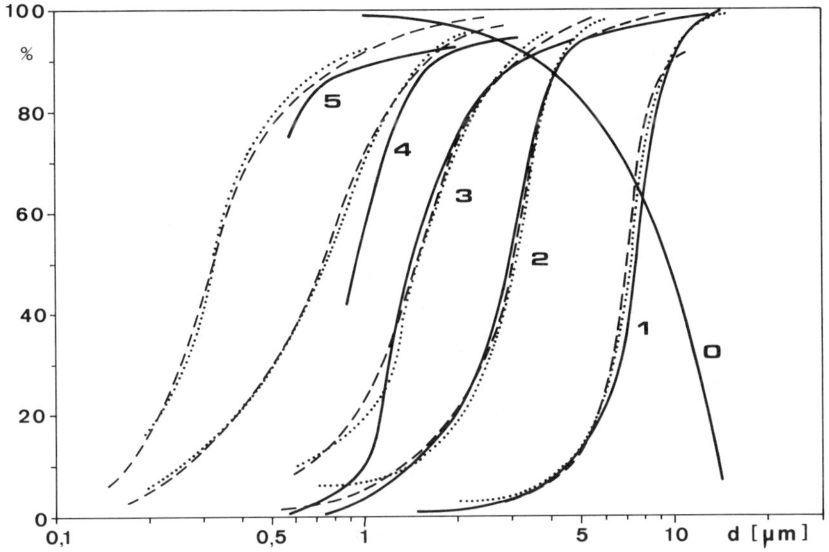

Fig. 5.3. Collection efficiency curves for each impactor stage as obtained with monodisperse particles. *Full curve* Manufacturer's specification. *Dashed curve* Literature data (Röbig et al. 1980). *Dotted curve* Fitting procedure used in the present study

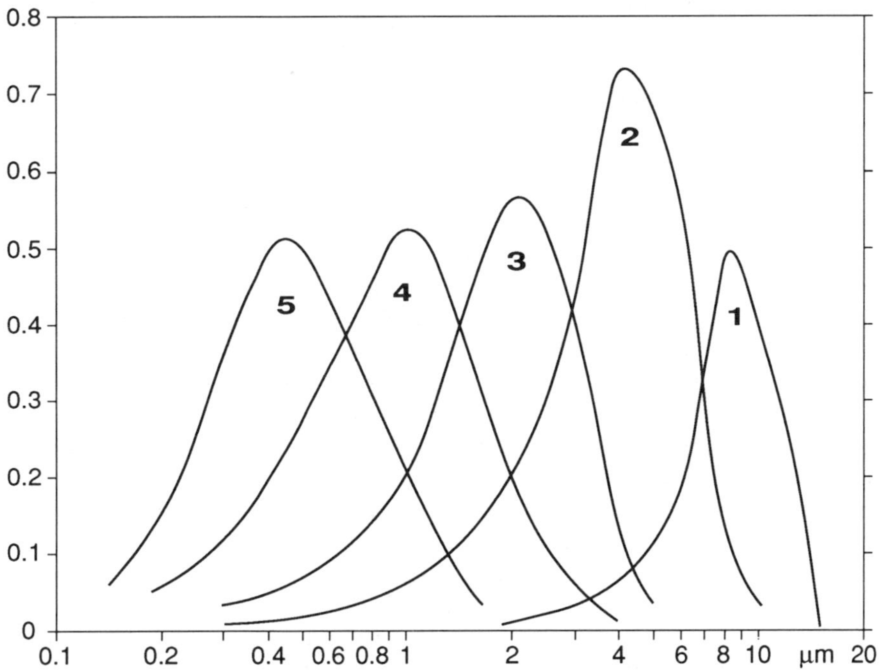

Fig. 5.4. Collection share of the individual impactor stages vs. equivalent aerodynamic particle diameter

Fig. 5.3 the contribution of the individual stages to the particle collection can be derived. The results are presented in Fig. 5.4.

5.1.4 Gas Measurements

To determine the dry deposition of gases the so-called gradient method was chosen, since for a broad spectrum of gases this is still the most promising technique (cf. Sect. 6.3). The vertical flux follows from the product of the turbulent exchange coefficient and the concentration gradient above the vegetation surface. The actual value of the exchange coefficient must be derived continuously from the micrometeorological data (Michaelis et al. 1988, 1992a; Schönburg and Mengelkamp 1989). The gas concentration has to be measured at least at three different heights above the forest canopy. As the trees were about 26 m high, air samples were taken at 48, 36, 28, 9 and 1 m above the ground. The last two heights were chosen in order to obtain additional information on processes within the ecosystem.

The gradient method makes high demands on the precision of the analytical results, since the concentration differences are often only in the 1 ppb order of magnitude. Though in itself the simultaneous operation of several analyzers for each gas species – one at each height – would be desirable, this procedure may cause difficulties due to effects such as zero drifts, nonlinearities or noise, not to mention that the financial expenditure involved is rather high. Therefore, another approach was chosen which used only one analyzer for each species. By means of a gas suction system with a computer-controlled valve device, air samples were taken successively from the various heights in a fixed timing rhythm and fed to the gas analyzers. In this way, vertical profiles of each species were obtained every 30 min. During this cycle after each change of the sampling height, the analytical values were not taken into account for an interval of 2.5 min, in order to minimize errors caused by memory effects and to make allowance for the final response time in the case of concentration jumps (cf. Sect. 5.2).

An important requirement on the suction system was to keep wall effects as small as possible. This could be achieved by taking several measures. The most important were the use of Teflon-coated pipes, a favourable ratio of volume to surface area and a short residence time of the air sample. The internal pipe diameter was 80 mm and the flow rate amounted to about 10 m/s. Thus, the sample stayed at most 6 s in the suction system. By automatically heating the outside insulated pipes to $\geq$10 K above ambient air temperature, condensation and absorption effects were further reduced.

Before mounting the pipe system onto the measuring tower at the investigation site, it was carefully tested in a laboratory experiment in order to quantify possible effects on the gas analysis and to derive appropriate correction functions. A schematic outline of these tests is given in Fig. 5.5. Air samples with a certain initial concentration were alternately sucked in via either a short unheated pipe (1.5 m) or a 45.1-m-long heatable pipe. In the

Table 5.1. Effect of a Teflon-coated pipe with 80 mm i. diam. on gas concentration measurements

Species		$\bar{t}_{air}$ [°C]		$\bar{V}_{pipe}$ [m/s]		Δp [h Pa]		$\bar{c}_k$ [b] [ppb]	$\bar{c}_l$ [ppb]	$\frac{\Sigma c_{l corr}}{n}$ [ppb]	Δc [ppb]	$(\Delta/c_k) \cdot 100$ [%]	$\frac{(\Delta c / c_k) \cdot 100}{\Delta l}$ [% / m]
		k[a]	l	k	l	k	l						$\Delta l = 43.6$ m
O_3		16.0	16.0	11.03	11.04	-5.50	-5.70	19.34	18.85		-0.49	-2.53	-0.058
O_3		16.0	16.0					11.40	11.06		-0.34	-2.98	-0.068
O_3	(Nat.)	16.5	17.0					24.28	23.97		-0.31	-1.28	-0.029
O_3		19.0	19.0					29.64	29.13		-0.51	-1.72	-0.039
NO	(Test gas)	17.0	17.0	10.83	10.82	-5.50	-5.70	46.02	45.07	45.02	-1.00	-2.17	-0.050
NO_x	(Test gas)	15.0	15.0	10.86	10.81	-5.50	-5.70	39.21	38.10	38.28	-0.93	-2.37	-0.054
SO_2	(Test gas)	15.0	15.0	10.90	10.94	-5.55	-5.70	17.48	17.22	17.28	-0.205	-1.17	-0.027
CO_2	(Nat.)	17.5	18.0	10.74	10.74	-5.50	-5.70	398.1 ppm	398.8 ppm		+0.70	< ±0.20	< ±0.005
O_3		13.0	36.0	10.71	10.88	-5.20	-5.45	17.19	16.74		-0.45	-2.62	-0.060
O_3	(Nat.)	11.7	35.4	10.76	10.92	-5.40	-5.55	18.58	18.30		-0.28	-1.51	-0.035
O_3		13.2	32.6			-5.45	-5.60	26.40	25.93		-0.47	-1.79	-0.041
NO	(Test gas)	16.0	34.0	10.80	10.83	-5.30	-5.40	37.27	39.01	36.68	-0.59	-1.58	-0.036
NO_x	(Nat.)	14.8	35.0	10.72	10.94	-5.40	-5.70	15.14	15.00		-0.14	-0.92	-0.021
NO_x	(Test gas)	16.0	34.3	10.71	10.92	-5.40	-5.60	33.03	34.30	32.75	-0.28	-0.82	-0.019
NO_x	(Nat.)	12.7	32.9	10.75	10.88	-5.45	-5.70	17.29	17.16		-0.13	-0.75	-0.017
SO_2	(Test gas)	15.9	35.3	10.79	10.91	-5.55	-5.70	18.32	19.24	18.23	-0.09	-0.49	-0.011
CO_2	(Nat.)	16.5	34.5	10.80	10.80	-5.30	-5.40	390.43 ppm	389.76 ppm		-0.67 ppm	< ±0.20	< ±0.005

[a] k, Unheated pipe of 1.5 m length; l, heatable pipe of 45.2 m length.
[b] Average of 60 to 200 single values each.

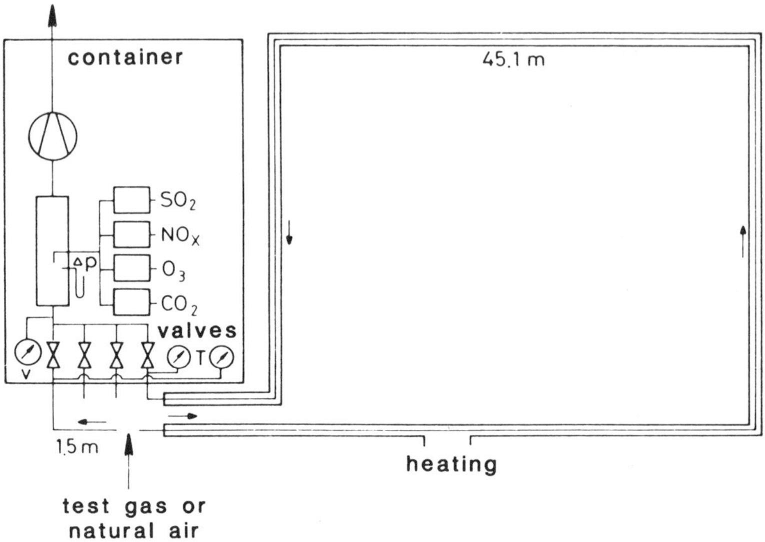

Fig. 5.5. Schematic setup for the tests of the air sampling system

case of O_3 and CO_2 as well as partially of NO_x, the measurements could be performed with natural ambient air, partly under different solar irradiation conditions. Otherwise the concentration was exactly adjusted in a dynamic mixing process by the addition of definite flows of test gases with analysis certificates. In this case, the raw concentration values measured after passage through the long pipe had to be corrected for the influence of temperature and flow rate. The results of the test experiments are summarized in Table 5.1. More details may be found elsewhere (Michaelis 1988; Michaelis et al. 1988).

As can be seen, the measured effects are rather small so that a linear dependence of the concentration on the pipe length may be assumed and a function of the form $(\Delta c/c_s)$ 100/l in [% /m] can be defined. Here Δc is negative and describes the concentration difference found behind the two pipes and c_s is the concentration at the exit of the short pipe. In the last column of Table 5.1, the values of this function are listed taking the path length difference in the experiment as a basis. As follows from these data, the necessary correction for the heated pipes is +0.045%/m in the case of O_3, +0.036%/m for NO, +0.019%/m for NO_x and +0.011%/m for SO_2. In the case of CO_2, as expected, no detectable falsification occurs. Thus, it can be concluded that the design of the sampling system described in general ensured an upper limit for the correction of about 1 ppb. This value was exceeded only during episodes with extreme concentrations, in particular of O_3 and at the 48-m inlet. In order to continuously control the functioning of the gas measurement system, a data trunk line to the institute at Geesthacht was installed.

5.1.5 Sampling of Soil Solution and Plant Components

Low-pressure suction candles (lysimeters) were used to extract soil solution near the measuring tower. In order to obtain results as representative as possible, four sets of these devices were positioned in suitable areas, with each set consisting of five lysimeters which took samples from 5, 15, 25, 45 and 165 cm depth. The soil water passed through a ceramic lysimeter head, a polyethylene suction pipe and connecting hose into a 2-l collecting reservoir. A pump ensured a low pressure of 0.5 bar for all devices. The methodology for determining fluxes will be described in Section 6.4. In general, weekly samples were taken (Rademacher et al. 1992).

For sampling plant material two principal methods were used. In individual cases, old spruce trees were pulled down with a tractor so that not only samples of needles, boughs and stem segments, but also of coarse and fine roots with the surrounding soil could be taken. On the other hand, small quantities of plant material were cautiously sampled from living trees, partly with the help of plastic-coated instruments. The latter method was used in particular for exploring seasonal variations, in the course of which sampling was carried out either monthly or quarterly (Rademacher et al. 1992).

5.2 Analytical Techniques

5.2.1 General Aspects

Though the accumulated input of atmospheric pollutants into terrestrial ecosystems often gives rise to grave concern, the concentration in samples within the scope of a comprehensive study can be very low and often down to the ultratrace level. Since the methodologies, which have to be applied, e.g., for deriving fluxes, already imply sources of errors (cf. Chap. 6), a high quality of the analytical data has to be assured. This requires sample handling free of losses and contamination, the application of well-approved sample preparation procedures and systematic intercomparisons of different analytical techniques. For instance, trace element determination in the present study was performed using total-reflection X-ray fluorescence (TXRF) and inductively coupled plasma optical emission spectroscopy (ICP-OES). Both methods were repeatedly checked against each other, and intercomparisons with other techniques such as atomic absorption spectroscopy (AAS), instrumental neutron activation analysis (INAA), differential pulse absorption voltammetry (DPAV) and differential pulse anodic stripping voltammetry (DPASV) were also carried out (Nürnberg et al. 1984; Michaelis et al. 1985a,c; Michaelis 1986b; Michaelis and Prange 1988; Pepelnik et al. 1994). Instrumental neutron activation analysis is of particular in-

Table 5.2. Results for a rainwater sample as obtained by different analytical techniques

	Element concentration (ng/g) TXRF						DPASV/DPAV[a]	
	Direct		Freeze-drying		Reverse-phase			
Element	Mean	r.s.d. (%)	Mean	r.s.d. (%)	Mean	r.s.d. (%)	Mean	r.s.d. (%)
S	1311	5.0						
K	244.5	4.3						
Ca	289.5	0.2						
Ti			0.41	0.3				
V			0.55	2.5	0.42	9.0		
Cr			< 0.1					
Mn	0.67	2.9	0.68	10.1	0.54	6.5		
Fe	3.14	6.1	3.26	7.6	2.91	2.7		
Co			Internal standard				0.04	3.0
Ni	0.52	13.2	0.43	8.5	0.43	2.1	0.43	6.5
Cu	0.50	17.3	0.39	7.5	0.52	5.0	0.36	15.7
Zn	3.20	8.8	3.51	5.3	3.14	12.6	3.45	1.0
As			0.12	8.0				
Se			0.12	14.1	0.12	5.3		
Rb			0.27	15.6				
Sr	4.70	3.0	4.84	6.6				
Mo			0.04	8.4	0.04	7.9		
Cd			0.12	11.6	0.10	12.3	0.08	7.6
Ba			1.0	4.6				
Pb	2.70	12.2	2.14	1.5	1.81	5.8	2.29	8.9

[a] DPASV, Differential pulse anodic stripping voltammetry; DPAV, differential pulse absorption voltammetry; r.s.d., relative standard deviation.

terest as a reference method, since no substantial sample preparation is required and, as a consequence, the risk of losses or contamination is rather low.

In order to illustrate realistic concentrations and relative standard deviations, the results for a rainwater sample as obtained by different analytical techniques are shown in Table 5.2 (Michaelis and Prange 1988). All data are given in ng/g. The diverse sample preparation methods for TXRF will be discussed in the next section. Generally, the errors in trace analysis are markedly underestimated by many authors. This is the definite outcome of numerous interlaboratory tests (Michaelis et al. 1985a; Michaelis and Prange 1988; Prange et al. 1993). It was an explicit objective of the present study to assure a high quality of the analytical results both with respect to precision and accuracy.

5.2.2 Sample Preparation for Trace Element Analysis

Both TXRF and ICP-OES analyses require liquid samples. This can be achieved by a procedure which was originally developed for TXRF, but which is also applicable to ICP-OES. The general preparation scheme is pre-

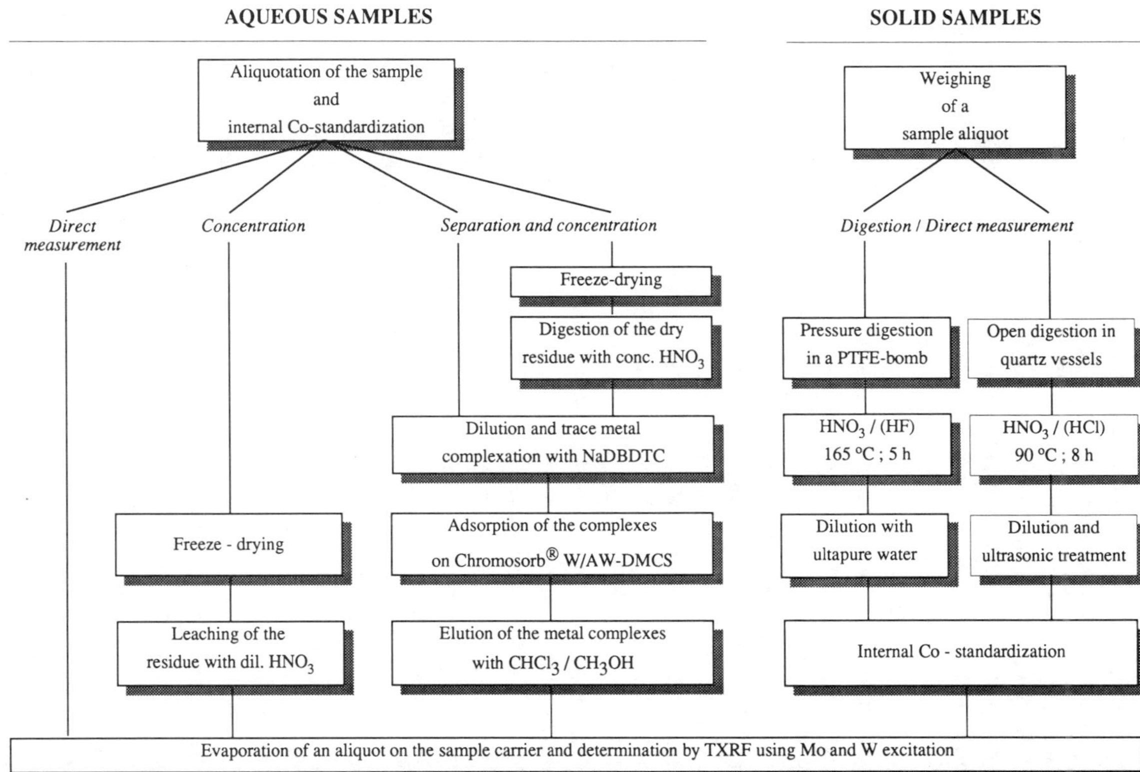

Fig. 5.6. General sample preparation scheme

sented in Fig. 5.6 (Stößel and Prange 1985; Prange et al. 1985; Michaelis and Prange 1988; Michaelis 1991; Prange and Schwenke 1992).

For aqueous solutions, sample preparation is often rather simple. Frequently, a direct measurement is possible. For instance, in the case of rainwater analysis with TXRF, the elements S, K, Ca, Mn, Fe, Ni, Cu, Zn, Sr and Pb can be detected in this way (cf. Table 5.2). Twenty to 100 μl of a Co standard solution are added to 20 ml of rainwater. After mixing, an aliquot of 25 μl is transferred to the siliconized sample carrier and vacuum-dried (cf. Sect. 5.2.3). For ICP-OES, part of the standardized solution is fed into a pneumatic nebulizer (cf. Sect. 5.2.4).

In other applications preconcentration and/or separation procedures are necessary. For aqueous samples, along with the direct measurement, freeze-drying and the reverse-phase technique are reliable and approved procedures. In the first case, 20 ml of the acidified sample, spiked with the Co standard, are freeze-dried. The residue is taken up in 1 ml of dilute (1:2) nitric acid (Merck, Suprapur) and leached for 2 h at 85 °C. A 25-μl aliquot is then dried on the sample support. The elements Ti, V, Cr, Mn, Fe, Co (if using another standard), Ni, Cu, Zn, As, Se, Rb, Sr, Mo, Cd, Ba and Pb can be detected using this technique. However, preconcentration by freeze-drying is only useful if the content of alkaline and alkaline-earth metals is low. Otherwise these latter elements will tend to enhance matrix effects due to their enrichment on the sample carrier.

In the case of very low trace element contents in the presence of marked alkaline and alkaline-earth metal concentrations, it is recommended to apply the reverse-phase technique which was originally developed for trace analysis of seawater (Prange et al. 1985). Two hundred ml of the acidified sample are spiked with an internal standard (e.g. 2–5 ng/g Se). The pH value is adjusted to range from 4.5 to 5 by means of a sodium hydroxide solution and 1 to 2 ml of acetate buffer (both Merck, Suprapur). Then 1 ml of a 4% (w/v) methanolic solution of sodium-dibenzyldithiocarbamate (NaDBDTC, Fluka) is added. The well-homogenized mixture is sucked through a pretreated reverse-phase column (Chromosorb AW-DMCS, Merck) within 10 to 15 min, allowing the carbamate complexes to be adsorbed. After the column has been sucked dry, the complexes are eluted with 3 to 4 ml of purified chloroform/methanol (70:30 v/v). Two hundred μl of the eluate are then evaporated in a special device on the sample carrier. In rainwater, the elements V, Mn, Fe, Ni, Cu, Zn, Se, Mo, Cd and Pb can be determined. For natural waters with higher concentrations of organic matter, this method is, however, not directly applicable. Reactions with natural complexing reagents and hydrophobic substances may affect the complex formation and the adsorption of the carbamate complexes on the column. In such cases, a preliminary digestion is required (Fig. 5.6).

For solid samples, such as soil or airborne particulates, digestion of the sample is applied. The preferable procedure is pressure digestion of, for instance, the loaded filters with 65% nitric acid in Teflon bombs at 165 °C for 5 to 6 h. The solution is then diluted with ultrapure water. Again, an internal standard is used. A total of 25 elements have been determined by TXRF using twin anode excitation (cf. Sect. 5.2.3): S, K, Ca, Ti, V, Cr, Mn, Fe, Ni, Cu, Zn, As, Se, Rb, Sr, Y, Zr, Nb, Mo, Ag, Cd, Sn, Sb, Ba and Pb. A typical example for organic matrices is the analysis of fine roots which are usually obtained with masses ranging from 40 to 160 µg. Samples are prepared by digesting the carefully washed fine roots with 100 µl conc. nitric acid at a temperature of about 70 °C. The acid is previously spiked with a Co standard solution.

5.2.3 Total-Reflection X-Ray Fluorescence (TXRF)

X-ray fluorescence analysis is a widely adopted analytical technique for measuring elemental concentrations. The particular advantages in comparison with other methods are simplicity in instrumentation, rapidity of measurement and the high degree of automation attainable. However, for trace analysis, the conventional method exhibits insufficient detection limits. This disadvantage is cancelled by the rather new total reflection X-ray fluorescence technique which makes use of the fact that the refractive index n for X-rays is slightly smaller than 1 (n=0.99999851 for the 17.5 keV K_α-radiation of molybdenum and quartz glass with a density of 2.2 g/cm^3). This leads to a critical angle of 5.9 min of arc for total reflection. If the exciting radiation strikes the optically flat quartz glass surface at glancing angles below this value, coherent and incoherent scattering occurs only within a very thin layer of less than 100 nm. At angles used in practice this depth even decreases to about 4 nm.

A schematic view of the instrument is shown in Fig. 5.7. The sample is prepared as a thin film on the quartz glass sample carrier. Since the exciting radiation only slightly enters the surface of the substratum, the associated reduction of background under the peaks in the fluorescence spectrum results in a substantial improvement of the detection limits compared to the conventional X-ray fluorescence technique. A second reflector in front of the sample carrier serves as a high-energy cutoff filter. In the form of a mirror or multilayer, it ensures that only radiation which fulfills the total-reflection condition reaches the sample support. Twin excitation from molybdenum

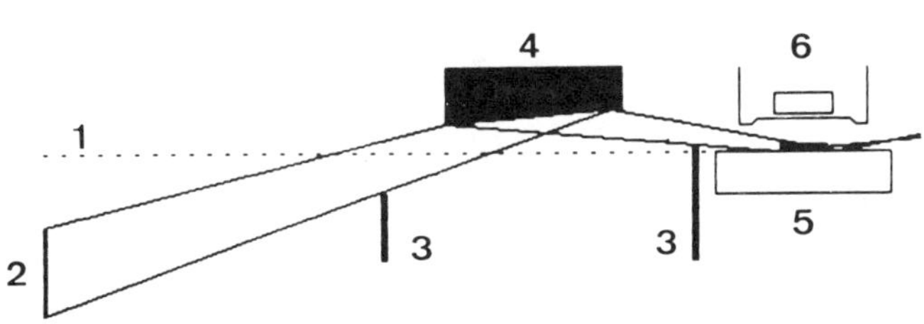

Fig. 5.7. Measurement principle of TXRF. *1* Reference plane; *2* anode of X-ray tube; *3* diaphragm; *4* reflector; *5* sample carrier with sample film; *6* detector

and tungsten anodes of a fine focus X-ray tube has been state of the art since the middle of the 1980s. The fluorescence radiation is detected by an Si(Li)-detector. Figure 5.8 shows the interference-free detection limits as obtained by means of mono-element aqueous solutions. The sensitivity is better than 10 pg for more than 60 elements (Prange and Schwenke 1992). In the case of realistic matrices and sampling procedures, the detection limits may be markedly higher. For instance, if weekly samples of airborne particulates are taken by means of a high-volume sampler with a five-stage impactor (cf. Sect. 5.1), then the detection limits for the differential concentrations in air range from 2 to 20 pg/m^3 (Michaelis and Prange 1988). A typical fluorescence spectrum of an aerosol sample is displayed in Fig. 5.9. The concurrence of pronounced multielement characteristics and high sensitivity makes TXRF increasingly attractive for trace element analysis.

The possibility of improving the signal-to-background ratio by total reflection of the exciting beam at a flat sample support was first pointed out in the early 1970s (Yoneda and Horiuchi 1971). The detection limits were quoted then to be in the order of a few nanograms. Some years later, the physical fundamentals were investigated in more detail (Aiginger and Wobrauschek 1974; Wobrauschek and Aiginger 1975). At that time all experimental systems were based on the use of optical benches. Because of the required precision in the adjustment of the components, the application remained restricted to individual studies. The efficiency was still rather poor. A decisive step towards the introduction of the method to routine analytical practice was achieved in the late 1970s by the successful design of a compact, stable and easily adjustable total-reflection module (Knoth and

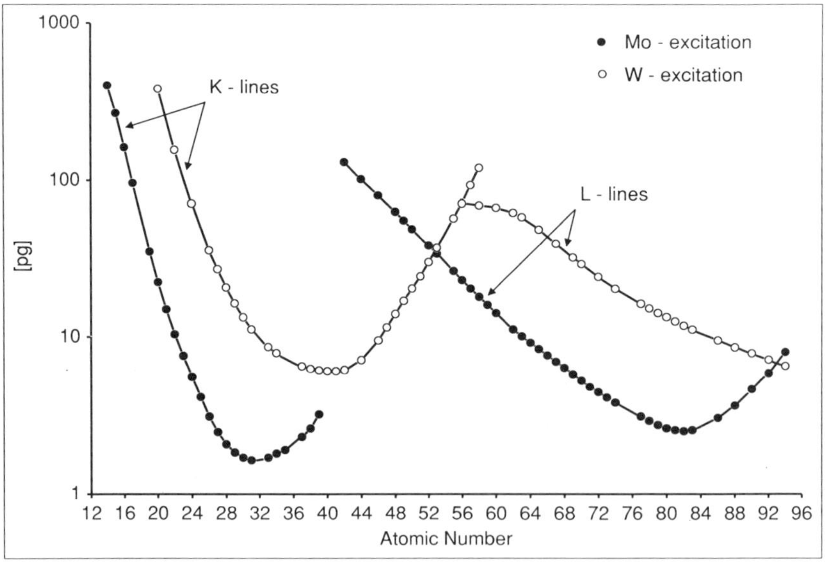

Fig. 5.8. Interference-free detection limits of TXRF in pg

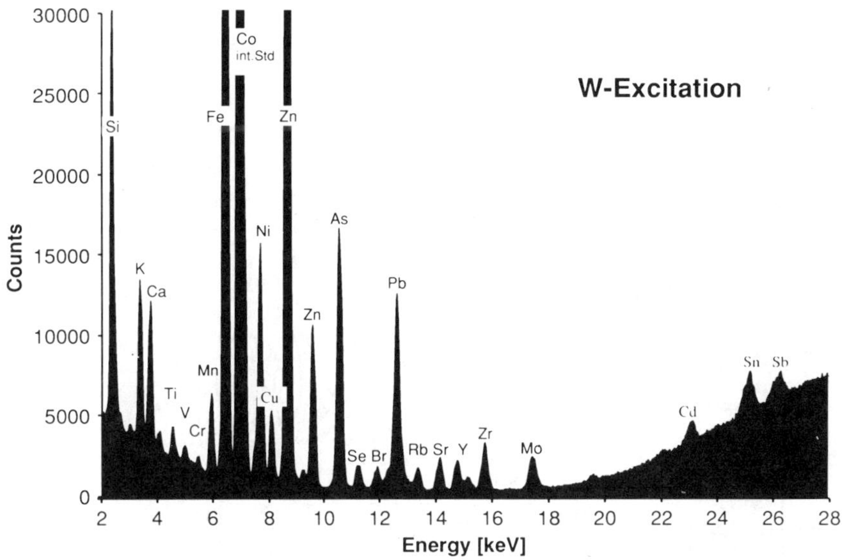

Fig. 5.9. X-ray fluorescence spectrum of an aerosol sample as obtained with tungsten anode excitation

Schwenke 1977, 1978, 1980; Schwenke et al. 1980; Schwenke and Knoth 1982). Over the following years, the technique was further improved, resulting in a highly efficient trace analytical tool (Prange and Schwenke 1992). Meanwhile TXRF has found a wide range of applications. Numerous studies have been reported from diverse disciplines, including estuarine and marine water quality management and research, mineralogical investigations, material science and surface analysis, archaeology, geochemistry, biology, biochemistry, medicine and forensic science (see, e.g., the following references and the literature cited therein: Michaelis et al. 1985a, 1985b, 1985c; Prange et al. 1985; Michaelis 1986b; Michaelis and Prange 1986, 1988; Prange et al. 1989, 1991; Klockenkämper et al. 1992; Michaelis et al. 1992b; Prange and Schwenke 1992; Pepelnik et al. 1993, 1994; Prange 1993; Prange et al. 1993, 1995). The spectrometer used in the present study (Extra II, Rich. Seifert & Co., Ahrensburg) was manufactured under licence from the GKSS Research Centre (Knoth and Schwenke 1977).

5.2.4 Inductively Coupled Plasma Optical Emission Spectroscopy (ICP-OES)

Optical emission spectroscopy with plasma excitation was the second basic technique for trace element analysis. It was applied, above all, in two major respects: (1) quality assurance of the analytical data (Michaelis 1986b) and (2) extension of the element spectrum towards low atomic numbers where

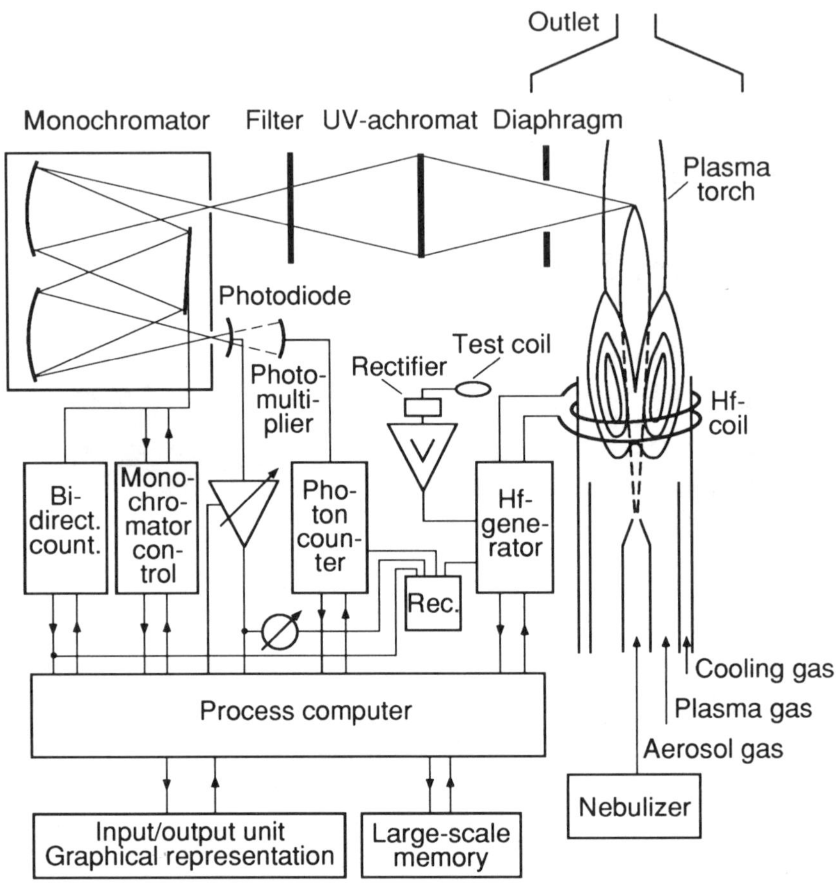

Fig. 5.10. Schematic setup of the sequential mode ICP-OES spectrometer

the quantum efficiency in X-ray analysis is rather poor. The following elements have been determined: Na, Mg, Al, Si, P, S, K, Ca, Mn, Fe, Cu, Zn, Sr, Ba and Pb. Samples were prepared as described in Section 5.2.2. Pneumatic atomization was used for feeding the sample into the plasma. Compared to ultrasonic atomizing this method offers several advantages, the most important of which being lower matrix effects (Boumans and DeBoer 1976), a better stability of the spray rate as well as a minor proneness to trouble (Bauer 1981).

Two spectrometers were available for the present study. The first one was a self-produced device operating in sequential mode (Fig. 5.10). All components had been systematically optimized in order to achieve outstanding performance with regard to accuracy and detection power in ultratrace analysis (Bauer 1981; Berneike 1987; Schönburg 1987). The second instrument was a simultaneous mode industrial product (ARL 34 000, Applied Research Lab., Sunland) in Rowland geometry with 32 photomultipliers. These two types of spectrometers are ideally complementary. The sequential mode device allowed a computer-controlled optimization of the parameters radio-frequency power, observation height in the plasma torch and aerosol gas flow for each spectral line during the measurements. As a consequence, the spectrometer is characterized by extremely low detection limits, down to markedly less than 1 ppb. Moreover, the possibility of resolving complex structures and of performing a realistic background correction in the spectra by means of a computer fit programme provides high accuracy of the analytical results. A typical example for such an analysis with Lorentz line profiles is shown in Fig. 5.11. On the other hand, the consumption of sample solution is rather high in the sequential mode so that quality assurance with the detection of few elements was in the foreground of the application. For routine measurements with large numbers of samples the simultaneous mode spectrometer has clear advantages because of the shorter measuring time and the lower sample mass required.

5.2.5 Ion Chromatography and pH Measurements

Besides trace element concentrations, ionic species in liquid samples represent further relevant environmental criteria. For the determination of SO_4^{2-}, NO_3^-, and Cl^- an ion-exchange chromatograph (2000i, Dionex, Idstein; separation column AS 3) was employed. NH_4^+ was detected using flow-injection analysis in combination with a gas-sensitive electrode (Perstorp Analytical, Rodgau). During a limited period (April 1987 to August 1989), F^- was also analysed by applying an electrode which consisted of a single-crystal lanthanum fluoride membrane and an internal reference (Orion, Cambridge,

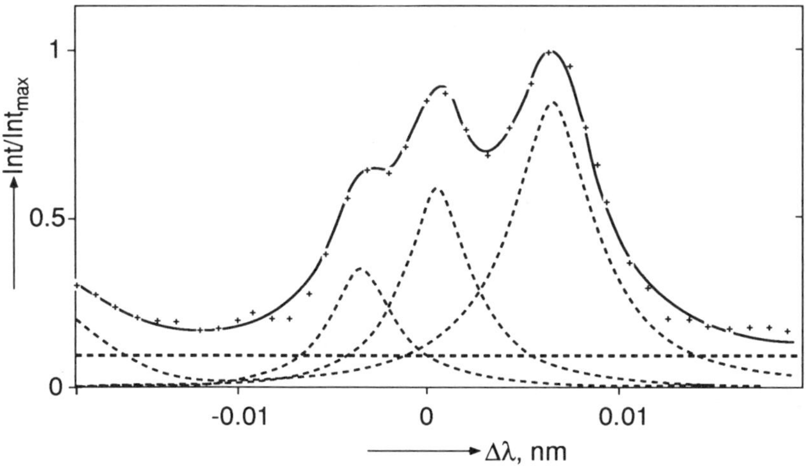

Fig. 5.11. Computer analysis of a complex structure in an OES-spectrum using Lorentz profiles

Massachusetts). pH measurements were performed using a microprocessor pH meter (Knick Elektron. Messgeräte, Berlin).

5.2.6 Gas Analysis

The gases NO and NO_2 were measured using a two-channnel chemiluminescence analyzer (Monitor Labs 8840, San Diego) which utilizes the reaction with O_3 from an ozone generator. Since only NO shows luminescence, this species is detected in the first channel. The second channel contains a molybdenum catalytic converter which at a temperature of approximately 315 °C reduces NO_2 to NO so that the sum of NO and NO_2, i.e. NO_x, is determined in this channel. The concentration of NO_2 is derived from the difference of the two output signals. With slight modifications the same device can be used to measure NH_3: a thermal converter added to the NO_x-channel transforms NH_3 to NO at a temperature of about 750 °C and the NO channel is supplemented by a molybdenum catalytic converter at 315 °C. In this way, on the one hand, the sum of NH_3 and NO_x and, on the other hand, NO_x are measured.

SO_2 was determined by UV-excited fluorescence (Monitor Labs 8850, San Diego). Metal vapour lamps (Cd 229 nm, Zn 214 nm) are suitable light

Table 5.3. Specifications of the gas analysers

Species	O_3	SO_2	NO/NO_2	CO_2	NH_3
Manufacturer	Thermo Electron	Monitor Labs	Monitor Labs	Hartmann & Braun	Monitor Labs
Model	TE 49	ML 8850	ML 8840	Uras 3E, 2T	ML 8840/8750
Measuring principle	UV absorption at 254 nm	UV fluorescence at 240–420 nm	Chemiluminescence at 500–3000 nm	IR absorption at 4 μm	Conversion to NO, chemiluminescence
Precision	2 ppb	2.5–5 ppb	~ 1 ppb at 100 ppb	2–4 ppm	~ 1 ppb
Linearity	± 1 ppb	± 1 % of measuring range	± 1 % of measuring range	± 1 %	± 1 % of measuring range
Detection limit	2 ppb	1 ppb	2 ppb	≤ 3 % of measuring range	2 ppb
Response time 0 to 95%	20 s	240 s	3 min	~ 2 min	20 min
Zero drift	< 0.5%/month	< 2 ppb/day	< 0.5%/week	≤ 1%/week	< 0.5%/week
Air throughput	1–3 l/min	500 ml/min	500 ml/min	500 ml/min	500 ml/min
Permissible ambient temperature	20–30°C	20–30°C	20–30°C	5–45°C	20-30°C

sources. O_3 was measured using UV photometry in a two-channel arrangement (Thermo-Electron 49, Hopkinton). In this device the air flow is alternately conducted through two absorption cells, one of which is preceded by a catalytic converter which quantitatively reduces O_3 in the sample. This measurement principle allows the elimination of the most important sources of error such as amplitude variations of the light source, interferences from other constituents absorbing in the UV, and light attenuation by components of the analyzer. The concentration of CO_2 was determined by infrared absorption at a wavelength of ~4 μm (Uras 2T and 3E, Hartmann & Braun, Frankfurt). The principle of partial layer detection in a two-channel arrangement with both measuring and reference cells ensures, to a large extent, corrections for overlapping absorption by other atmospheric gases.

The most important specifications of the analyzers are summarized in Table 5.3. A particular comment should be made regarding NH_3. The long time constant in this case excludes deposition measurements when only one device is employed for determining the concentration gradient (cf. Sect. 5.1.4). Regularly once a week each analyzer was carefully calibrated using certified test gases and standard procedures.

References

Aiginger H, Wobrauschek P (1974) A method for quantitative X-ray fluorescence analysis in the nanogram region. Nucl Instrum Methods 114:157–158

Bauer M (1981) Untersuchungen zur optischen Emissionsspektroskopie mit induktiv angeregter Plasmafackel. Thesis, University of Hamburg. GKSS Forschungszentrum Geesthacht, GKSS 82/E/9

Berneike W (1987) Untersuchungen zu Einsatzmöglichkeiten der optischen Emissionsspektrometrie mit induktiv geheizter Argonplasmafackel bei der Analyse von Umweltproben. Thesis, University of Hamburg. GKSS Forschungszentrum Geesthacht, GKSS 87/E/36

Boumans PWJM, DeBoer FJ (1976) Studies of a radio frequency inductively coupled argon plasma for optical emission spectrometry. III. Interference effects under compromise conditions for simultaneous multi-element analysis. Spectrochim Acta 31B:355–375

Dannecker W, Naumann K, Bergmann J (1982) Untersuchung der Korngrößenverteilung von Schwebstäuben sowie des Eluationsverhaltens darin enthaltener umweltrelevanter Elemente. Staub–Reinhalt Luft 42,4:176–182

Klockenkämper R, Knoth J, Prange A, Schwenke H (1992) Total-reflection X-ray fluorescence spectroscopy. Anal Chem 64(23):1115–1121

Knoth J, Schwenke H (1977) GKSS Patent P 27 36 960.4

Knoth J, Schwenke H (1978) An X-ray fluorescence spectrometer with totally reflecting sample support for trace analysis at the ppb level. Fresenius Z Anal Chem 291:200–204

Knoth J, Schwenke H (1980) A new totally reflecting X-ray fluorescence spectrometer with detection limits below 10^{-11} g. Fresenius Z Anal Chem 301:7–9

Michaelis W (1986a) Naß- und Trockendeposition von Schwermetallen. In: Klose W, Leßmann E (eds) Bodenschutz – Lösung durch Technik, ENVITEC 86. Vulkan-Verlag, Essen, pp 36–41

Michaelis W (1986b) Multielement analysis of environmental samples by total-reflection X-ray fluorescence spectrometry, neutron activation analysis and inductively coupled plasma optical emission spectroscopy. Fresenius Z Anal Chem 324:662–671

Michaelis W (1988) Experimental studies on dry deposition of heavy metals and gases. In: van Dop H (ed) Air pollution modeling and its application VI, vol 11. Plenum Press, New York, pp 61–74

Michaelis W (1991) Trace analysis of geological and environmental samples. In: Störr M, Henning KH, Adolphi P (eds) Proceedings 7th EUROCLAY Conference, Dresden '91, vol II. Ernst-Moritz-Arndt Universität Greifswald, pp 761–766

Michaelis W, Prange A (eds)(1986) Totalreflexions-Röntgenfluoreszenzanalyse - 1st Workshop. GKSS Forschungszentrum Geesthacht, GKSS 86/E/61

Michaelis W, Prange A (1988) Trace analysis of geological and environmental samples by total-reflection X-ray fluorescence spectrometry. Nucl Geophys 2(4):231–245

Michaelis W, Fanger HU, Niedergesäß R, Schwenke H (1985a) Intercomparison of the multielement analytical methods TXRF, NAA and ICP with regard to trace element determinations in environmental samples. In: Sansoni B (ed) Instrumentelle Multielementanalyse. VCH Verlagsgesellschaft, Weinheim, pp 693–709

Michaelis W, Prange A, Knoth J (1985b) Applications of total-reflection X-ray fluorescence in multi-element analysis. In: Sansoni B (ed) Instrumentelle Multielementanalyse. VCH Verlagsgesellschaft, Weinheim, pp 269–289

Michaelis W, Knoth J, Prange A, Schwenke H (1985c) Trace analytical capabilities of total-reflection X-ray fluorescence analysis. In: Barret CS, Predecki PK, Leyden DE (eds) Advances in X-ray analysis, vol 28. Plenum Press, New York, pp 75–83

Michaelis W, Schönburg M, Stößel RP (1988) Trocken- und Naßdeposition von Schwermetallen und Gasen. In: Bauch J, Michaelis W (eds) Das Forschungsprogramm Waldschäden am Standort "Postturm", Forstamt Farchau/Ratzeburg. GKSS Forschungszentrum Geesthacht, GKSS 88/E/55, pp 19–59

Michaelis W, Schönburg M, Stößel RP (1989) Deposition of atmospheric pollutants into a North German forest ecosystem. In: Georgii HW (ed) Mechanisms and effects of pollutant-transfer into forests. Kluwer, Dordrecht, pp 3–12

Michaelis W, Pepelnik R, Theopold F, Rademacher P (1992a) Deposition atmosphärischer Spurenstoffe und Stoffflüsse im Ökosystem Wald. In: Michaelis W, Bauch J (eds) Luftverunreinigungen und Waldschäden am Standort "Postturm", Forstamt Farchau/Ratzeburg. GKSS Forschungszentrum Geesthacht, GKSS 92/E/100, pp 11–59

Michaelis W, Pepelnik R, Prange A (1992b) Application of TXRF in environmental research. In: Barret CS, Gilfrich JV, Huang TC, Jenkins R, McCarthy GJ, Predecki PK, Ryon R, Smith DK (eds) Advances in X-ray analysis, vol 35B. Plenum Press, New York, pp 953–958

Michaelis W, Pepelnik R, Rademacher P, Riebesell M (1992c) Transfer of atmospheric pollutants into a forest ecosystem. In: Teller A, Mathy P, Jeffers JNR (eds) Responses of forest ecosystems to environmental changes. Elsevier, London, pp 596–597

Nguyen VD, Valenta P (1978) KFA Patent 2831 8403

Nürnberg HW, Valenta P, Nguyen VD, Gödde M, Urano de Carvalho E (1984) Studies on the deposition of acid and of ecotoxic heavy metals with precipitates from the atmosphere. Fresenius Z Anal Chem 317:314–323

Panten A (1990) Analyse des Elementgehalts der Kronentraufe unter Fichten in einem geschädigten Waldgebiet. Diplomarbeit, Fachbereich Angewandte Naturwissenschaften, Fachhochschule Lübeck

Pepelnik R, Erbslöh B, Michaelis W, Prange A (1993) Determination of trace element deposition into a forest ecosystem using total-reflection X-ray fluorescence. Spectrochim Acta 48B(2):223–229

Pepelnik R, Prange A, Niedergesäß R (1994) Comparative study of multi-element determination using inductively coupled plasma mass spectrometry, total-reflection X-ray fluorescence spectrometry and neutron activation analysis. J Anal Atom Spectrom 9:1071–1074

Prange A (1993) Totalreflexions-Röntgenfluoreszenz. Nachr Chem Tech Lab 41(1):40–45

Prange A, Schwenke H (1992) Trace element analysis using total-reflection X-ray fluorescence spectrometry. In: Barret CS, Gilfrich JV, Huang TC, Jenkins R, McCarthy GJ, Predecki PK, Ryon R, Smith DK (eds) Advances in X-ray analysis, vol 35B. Plenum Press, New York, pp 899–923

Prange A, Knöchel A, Michaelis W (1985) Multi-element determination of dissolved heavy metal traces in sea-water by total-reflection X-ray fluorescence spectrometry. Anal Chim Acta 172:79–100

Prange A, Böddeker H, Michaelis W (1989) Multi-element determination of trace elements in whole blood and blood serum by TXRF. Fresenius Z Anal Chem 335:914–918

Prange A, Kramer K, Reus U (1991) Determination of trace element impurities in ultrapure reagents by total-reflection X-ray spectrometry. Spectrochim Acta 46B(10):1385–1393

Prange A, Böddeker H, Kramer K (1993) Determination of trace elements in river-water using total-reflection X-ray fluorescence. Spectrochim Acta 48B(2):207–215

Prange A, Reus U, Böddeker H, Fischer R, Adolf FP (1995) Microanalysis in forensic science: characterization of single textile fibers by total-reflection X-ray fluorescence. Anal Sci 11: 483–487

Rademacher P, Ulrich B, Michaelis W (1992) Bilanzierung der Elementvorräte und Elementflüsse innerhalb der Ökosystemkompartimente Krone, Stamm, Wurzel und Boden eines belasteten Fichtenbestandes am Standort "Postturm". In: Michaelis W, Bauch J (eds) Luftverunreinigungen und Waldschäden am Standort "Postturm", Forstamt Farchau/Ratzeburg. GKSS Forschungszentrum Geesthacht, GKSS 92/E/100, pp 149–186

Rao AK, Whitby KT (1978) Non-ideal collection characteristics of inertial impactors, parts I and II. J Aerosol Sci 9:77–100

Röbig G, Becker KH, Hessin A, Porstendörfer J, Scheibel HG (1980) A cascade impactor calibration for measurements of activity size distributions in the atmosphere. In: Gesellschaft für Aerosolforschung Schmallenberg (ed) Proc 8th Conf Aerosols in Science, Medicine and Technology, pp 96–102

Schönburg M (1987) Radiometrische Datierung und quantitative Elementbestimmung in Sediment-Tiefenprofilen mit Hilfe kernphysikalischer sowie Röntgenfluoreszenz- und atomemissionsspektrometrischer Verfahren. Thesis, University of Hamburg. GKSS Forschungszentrum Geesthacht, GKSS 87/E/54

Schönburg M, Mengelkamp HT (1989) Turbulente Flüsse in der atmosphärischen Grenzschicht über einem Waldgebiet. In: Arbeitsgemeinschaft der Großforschungseinrichtungen (AGF) (ed) Wechselwirkung Atmosphäre-Biosphäre. Thenée Druck, Bonn, pp 5–7

Schultz J (1991) Untersuchung von Traufenwasser und Bodenlösung auf Nähr- und Schadelemente - Versuch einer Bilanz. Diplomarbeit, Technische Universität Berlin

Schwenke H, Knoth J (1982) A highly sensitive energy-dispersive X-ray spectrometer with multiple total reflection of the exciting beam. Nucl Instr Methods 193:239–243

Schwenke H, Knoth J, Michaelis W (1980) Ultra-sensitive trace element analysis of environmental samples using advanced TRXRF techniques. In: Vogt JR (ed) Proc 4th Int Conf Nucl Methods Environ Energy Res, Columbia. CONF-800 433, GKSS 80/E/39, pp 313–320

Stößel RP (1986) Untersuchungen zur Naß- und Trockendeposition von Schwermetallen auf der Insel Pellworm. Thesis, University of Hamburg. GKSS Forschungszentrum Geesthacht, GKSS 87/E/34

Stößel RP, Prange A (1985) Determination of trace elements in rainwater by total-reflection X-ray fluorescence. Anal Chem 57:2880–2885

Winkler P (1974) Relative humidity and the adhesion of atmospheric particles on the plates of impactors. J Aerosol Sci 5:235–240

Winkler P (1985) Verfahren der Depositionsmessung. Staub–Reinh Luft 45, 6:256–260

Winkler P (1993a) Principles of automatic rain detection. Meteorol Z, NF 2:27–34

Winkler P (1993b) Response of precipitation sensors to rain. Meteorol Z, NF 2:35–44

Winkler P, Jobst S, Harder C (1989) Meteorologische Prüfung und Beurteilung von Sammelgeräten für die nasse Deposition. Gesellschaft für Strahlen- und Umweltforschung (GSF), Neuherberg, BPT-Bericht 1/89

Winkler P, Riedl J, Lang P (1993) A threshold intensity to standardize wet deposition. Meteorol Z, NF 2:21–26

Wobrauschek P, Aiginger H (1975) Total-reflection X-ray fluorescence spectrometric determination of elements in nanogram amounts. Anal Chem 47:852–855

Yoneda Y, Horiuchi T (1971) Optical flats for use in X-ray spectrochemical microanalysis. Rev Sci Instrum 42:1069–1070

6 Methodology for the Determination of Atmospheric Deposition and of Fluxes within the Ecosystem Compartments

6.1 Basic Considerations on Dry Deposition

To determine the dry deposition of atmospheric constituents several procedures have been described in the literature. Expenditure and chances of success are quite different and depend strongly on the attendant conditions, the constituent to be measured and the quality of the available technique. In any case, the quantification of dry deposition is much more difficult and subject to errors as compared to that of wet deposition.

A promising procedure for measuring fluxes of trace elements in aerosols is the so-called concentration method. If $c_{z_0 i}$ is the concentration measured at the reference height z_0 in the particle-size fraction i (cf. Sect. 5.1.3), and if v_{Di} is the corresponding deposition velocity, then the flux F to the surface follows from the equation

$$F = \sum_i c_{z_0 i} v_{Di} \,. \tag{1}$$

Besides correct size fractionation of airborne particulates, the central problem of this method lies in the degree of knowledge of the deposition velocity which strongly depends on the equivalent aerodynamic diameter and the particle surface properties. This will be discussed in more detail in the next section.

When using the so-called gradient method (Businger et al. 1971), the flux F is derived from the product of the concentration gradient $\Delta c/\Delta z$ and the vertical turbulent exchange coefficient K_z:

$$F = K_z \frac{\Delta c}{\Delta z} \tag{2}$$

by analogy to the well-known diffusion process. At least three measuring positions are arranged on a tower, preferably at logarithmically graded distances. In the case of gases, this can be rather easily realized by the use of appropriate suction pipes (cf. Sect. 5.1.4). Problems arise, however, when the deposition of trace elements by airborne particulates is to be determined. The demands on precision and accuracy in the analytics are very high, since in most cases the concentration differences between adjacent measuring positions range only from 1 to 10 % (Hicks and Wesely 1978). Thus, the uncertainties of the results for trace elements are often unacceptable. The method, however, may be of interest in special applications where the main constituents in aerosols are to be determined. An example is given in the subsequent section. Another complication, when applying the gradient method to particulates, is due to the fact that with increasing particle size the contribution of sedimentation rises so that Eq. (2) has to be supplemented by an additional term. In any case when using the gradient method, procedures have to be elaborated for ascertaining the current turbulent exchange coefficient (cf. Sect. 6.3).

According to the turbulence theory, the vertical flux can also be derived using the so-called eddy correlation method:

$$F = \overline{w'c'}, \tag{3}$$

where the temporal mean of the product of the turbulent components w' and c' of vertical wind velocity and concentration, respectively, has to be determined. It is obvious that this procedure requires sensors with short response times (>1 c/s). For a few gases this condition can be fulfilled by the use of optical methods, but for the combination of gases investigated in the present study the eddy correlation method was not considered sufficiently promising. For measuring w', the ultrasonic technique is feasible. In the case of trace element deposition, the experimental equipment becomes complex and demanding, and the analytics encounters grave difficulties.

The requirement of short response times for the concentration measurement can be circumvented by the 'turbulence accumulation' method which integrates the concentration over a period T (Hicks and Wesely 1978). A fast anemometer for the vertical wind component controls the loading of two filters in such a way that one of them is loaded only at upward-directed and the other only at downward-directed wind. The air throughput must in each case be with a known factor a proportional to the vertical wind velocity. When $m\uparrow$ and $m\downarrow$ are the respective filter loads, then the flux is given by

$$F = \frac{1}{aT}(m\uparrow - m\downarrow). \tag{4}$$

The mass differences are in the order of few percent so that this method also makes high demands on the analysis.

The use of surrogate surfaces made of chemically neutral materials such as Teflon or polyethylene appears to be a rather simple and, hence, particularly attractive approach. In principle, the amount of substances deposited may be determined by rinsing and analysing the wash solution. Shape and surface properties, however, deviate considerably from the characteristics of natural surfaces so that the results are very unsatisfactory (Dolske and Gatz 1984). This disadvantage may be avoided by using natural instead of artificial surfaces such as, for instance, leaves. Here, the problem lies in the washing procedure which must be quantitative without leaching out elements from the leaf interior. In general, these techniques suffer from the difficulty of obtaining representative samples.

Because of their simplicity sedimentation vessels, covered during precipitation events, are often used for monitoring the dry deposition of trace elements. The essential drawback of this method is that preferably larger particulates are collected which are subject to sedimentation. The sampling of smaller particles undergoing turbulent diffusion and Brownian motion remains incomplete. In this connection it is important to point out that just in the case of anthropogenic trace elements the concentration in air increases with decreasing aerodynamic diameter (cf. Chap. 9) and that the deposition velocity of small particles was for a long time underestimated. This point will be discussed in the next section.

On the basis of these considerations, it was decided in the present study to apply the concentration method to the measurement of the dry deposition of airborne particulates and the gradient method for determining the deposition of atmospheric gases. The fundamental aspects which have to be considered are treated in the following sections.

6.2 Aerosol Measurements

Equation (1) requires the knowledge of the particle size-dependent deposition velocity v_D with due regard to the nature of the vegetation and the current meteorological boundary conditions. v_D is a function of the friction velocity u_* and in a good approximation proportional to it. In the following, all data refer to the reference height z_o=1 m and a friction velocity u_* of about 30 cm/s.

Several studies have been performed to determine v_D for different applications. Within a semiempirical model (Sehmel and Hodgson 1974; Sehmel 1980), the flux which is controlled by various physical processes, including gravitational settling, turbulent diffusion and Brownian motion, was calculated from meteorological data and particle variables. The influence of the vegetation properties was derived from wind-tunnel experiments using different surface materials, particle diameters, friction velocities and roughness lengths. In field experiments, the two latter quantities must be deduced from measurements of the wind profile and other data (cf. Sect. 6.3).

Another approach consists in utilizing deposition velocities determined in field tests by means of labelled substances (Chamberlain 1966; Jonas and Vogt 1982; Jonas 1983, 1984). In Fig. 6.1, the results obtained with these two concepts are compared for the rather simple case of deposition onto a grass surface. It is evident that the semiempirical model predicts a minor dependence on the particle diameter in comparison to the field deposition experiments.

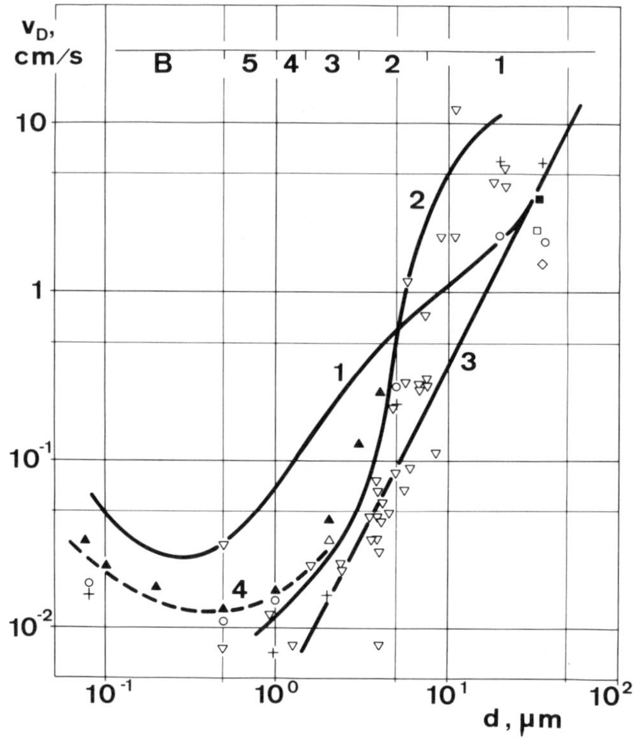

Fig. 6.1. Deposition velocity v_D of aerosols onto grass vs. particle diameter d. u_*=30 cm/s. Measured values from field and wind tunnel experiments (see text). *Curve 1* Prediction of the semiempirical model. Reference height 1 m, roughness length 3 cm, neutral atmospheric stability. *Curve 2* Summary of the Jülich deposition experiments. *Curve 3* Gravitational settling velocity. *Curve 4* Extrapolation of measured values to small particle diameters. *Above* The approximate collection intervals of the impactor are indicated. *B* Backup filter

Even more severe discrepancies occur in the case of deposition onto a spruce canopy. Obviously, the results of experimental studies on the deposition velocity strongly depend on the methods applied, as is shown in Fig. 6.2. During the initial stage of the present investigation, the findings obtained with artificial monodisperse, radiolabelled aerosols (Jonas and Vogt 1982; Jonas 1983, 1984) were taken as a basis for deriving the dry deposition of trace elements. However, very soon scanning electron microscope studies of natural particles deposited on needles (Waraghai and Gravenhorst 1989; Gravenhorst and Waraghai 1990) gave rise to serious doubts as to whether experiments using artificial particles correctly reflect the real conditions. This conclusion is supported by results derived from throughfall chemistry and radioactivity measurements after the Chernobyl accident (Höfken and Gravenhorst 1983; Grosch and Schmitt 1988; Brückmann 1988). Figure 6.2 clearly illustrates these findings which suggest a parallelism to the different behaviours of natural and artificial particles with respect to the impactor collection efficiencies (cf. Sect. 5.1.3).

In order to acquire further evidence of this phenomenon, independent investigations were carried out at the site of the present study (Hertlein 1990). They were based on the supposition that small particulates follow turbulent and molecular diffusion and thus the flux to the canopy can be determined by means of the gradient method. Then, by combining Eqs. (1) and (2) in a differential form, one obtains for the deposition velocity:

$$v_{Di} = \frac{K_z}{c_{z_0 i}} \frac{\Delta c_i}{\Delta z}. \quad (5)$$

The index i refers to the particle-size fraction. In order to ensure negligible contributions of gravitational settling, only the stage 5 and the backup filter of the impactors were evaluated. Trace analysis was restricted to those elements which can be detected with high precision and accuracy. These are S, Ca, Mn, Fe, Cu and Pb. The results have been included in Fig. 6.2. As can be

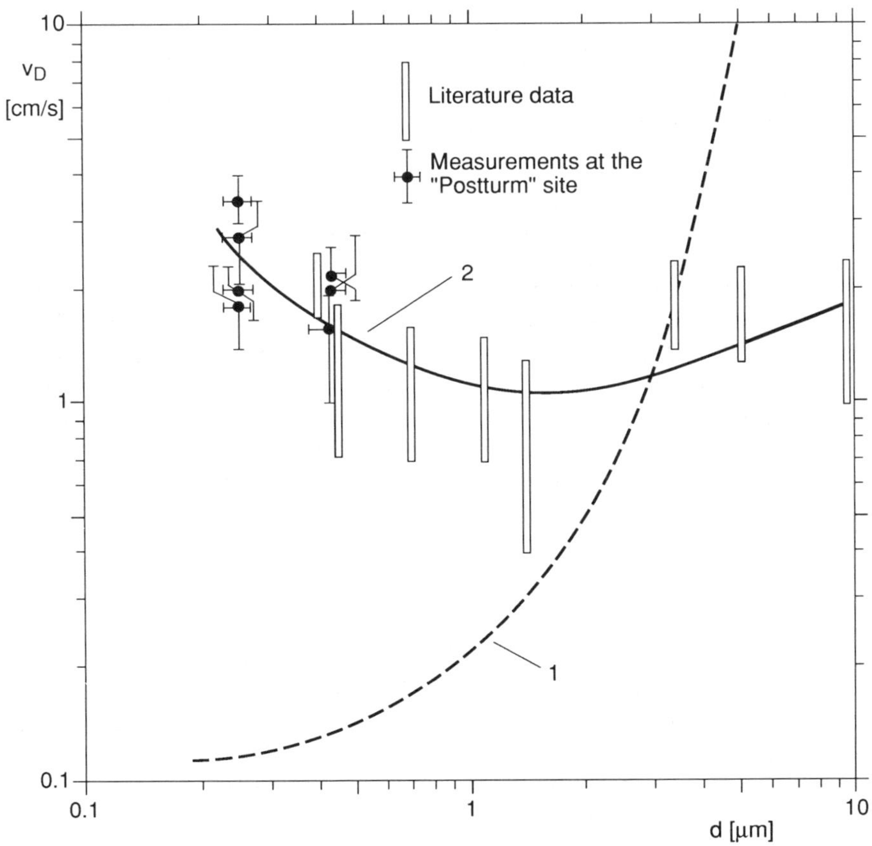

Fig. 6.2. Deposition velocities of airborne particulates onto a spruce stand. *Curve 1* Experiments with artificial monodisperse, radiolabelled aerosols. *Curve 2* Studies on the basis of natural particle deposition. See text

seen, there is excellent agreement with other studies on natural aereosols. Therefore, the corresponding fit was used in the present investigation. In comparison to the earlier data, the dependence on the aerodynamic particle diameter is markedly weaker. This is also predicted by the semiempirical model mentioned above. The effects on the derived fluxes vary from element to element according to the concentration distribution vs. particle size (cf. Chap. 9). A verification of the chosen procedure follows from the fact that the flux balances within the ecosystem exhibit a much better consistency (cf. Chap. 11).

6.3 Turbulent Fluxes in the Boundary Layer

When applying the gradient method (Businger et al. 1971) according to Eq. (2), it is assumed that the mechanisms of the turbulent transfer of pollutants are the same as those which also govern the exchange of momentum, sensible heat or humidity. The exchange coefficient K_C for an atmospheric constituent and those for momentum (K_M) and sensible heat or humidity (K_H) are set proportional to each other:

$$K_C \cong K_M \cong K_H. \tag{6}$$

Utilizing the flux-gradient relationship, K_M and K_H are determined for the horizontally homogeneous boundary layer. In order to assure equilibrium flow conditions at the measuring station, the distance to the edge of the forest should be at least 1 km (Tajchman 1981). At the "Postturm" site this demand was fulfilled for wind directions between north-northeast and south-southeast. From other directions the distance was less and down to about 300 m (see below).

In the lower boundary layer, the vertical turbulent exchange coefficient can be described by the product of the friction velocity u_* and the scale length $\kappa(z-d)$:

$$K_{M,H} = \frac{\kappa(z-d)\,u_*}{\Phi_{M,H}}. \tag{7}$$

Here, d denotes the so-called displacement height and $\kappa=0.4$ is the von Karman constant. The displacement height is introduced in order to extend the theory to terrain with high vegetation. It is assumed that $d_M=d_H$. The stability functions $\Phi_{M,\,H}$ make allowance for deviations in the wind and temperature profiles from the logarithmic shape in the case of non-neutral atmospheric stratification:

$$\Phi_M = \frac{\kappa(z-d)}{u_*}\frac{\delta u}{\delta z}, \tag{8}$$

$$\Phi_H = \frac{\kappa(z-d)}{R\Theta_*}\frac{\delta\Theta}{\delta z}. \tag{9}$$

The factor R in Eq. (9) takes into account possible differences in the exchange coefficients for momentum and sensible heat. Θ stands for the so-called potential temperature

$$\Theta = T\left(\frac{p_o}{p}\right)^k \tag{10}$$

with the absolute temperature T, the atmospheric pressure p, p_o=1000 hPa and k=0.286. Finally, Θ_* is the scale quantity for the temperature profile analogous to u_* for the wind profile. For the three cases of atmospheric stratification, the functions $\Phi_{M,\,H}$ can be represented as follows (Businger et al. 1971):

instable:

$$\Phi_M = \left(1-15\frac{z}{L}\right)^{-0.25} \qquad \Phi_H = \left(1-9\frac{z}{L}\right)^{-0.5} \tag{11}$$

neutral:

$$\Phi_M = \Phi_M = 1 \tag{12}$$

stable:

$$\Phi_M = \Phi_H = 1 + 4.7\frac{z}{L}. \tag{13}$$

L in these equations denotes the Monin-Obukhov length which is given by

$$L = \frac{\overline{T}u_*^2}{g\kappa\Theta_*}, \tag{14}$$

where $\overline{T}$ is the mean absolute temperature and g the gravitational acceleration.

Integration of the Eqs. (8) and (9) over the height z_1 to z_2 reveals:

$$u(z_2-d)-u(z_1-d) = \frac{u_*}{\kappa}\left[\ln\frac{z_2-d}{z_1-d} - \Psi_M\left(\frac{z_2-d}{L}\right) + \Psi_M\left(\frac{z_1-d}{L}\right)\right] \tag{15}$$

for the wind profile and

$$\Theta(z_2-d)-\Theta(z_1-d) = \frac{R\Theta_*}{\kappa}\left[\ln\frac{z_2-d}{z_1-d} - \Psi_H\left(\frac{z_2-d}{L}\right) + \Psi_H\left(\frac{z_1-d}{L}\right)\right] \tag{16}$$

for the temperature profile. The functions $\Psi_{M,H}$ were determined after Paulson (1970). In the case of instable stratification, one obtains:

$$\Psi_M = 2\ln\left(\frac{1+X}{2}\right) + \ln\left(\frac{1+X^2}{2}\right) - 2\arctan X + \frac{\pi}{2} \tag{17}$$

and

$$\Psi_H = 2\ln\left(\frac{1+Y}{2}\right) \tag{18}$$

with

$$X = \left(1 - 15\frac{z}{L}\right)^{0.25};\ Y = \left(1 - 9\frac{z}{L}\right)^{0.5}.$$

For stable conditions, the relations

$$\Psi_M = -4.7\frac{z}{L} \text{ and } \Psi_H = -\frac{4.7}{R}\frac{z}{L} \tag{19}$$

are valid. In the case of neutral stratification, both Ψ_M and Ψ_H are zero.

The displacement height d was determined from wind profiles under quasi-neutral conditions by ascertaining for each profile those values which delivered the best approximation to the logarithmic shape (Oliver 1971). Be-

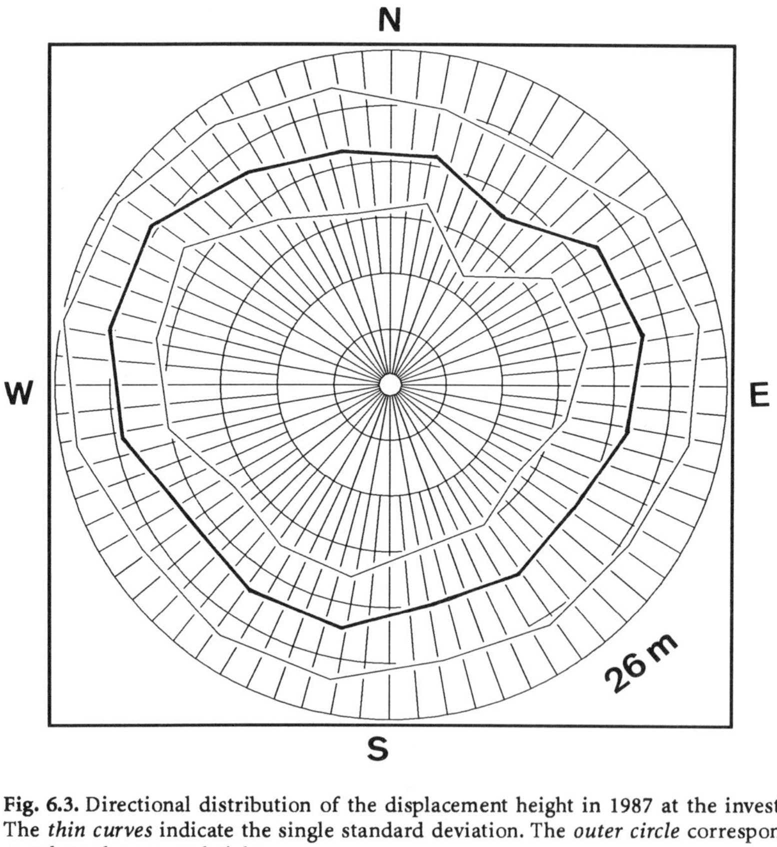

Fig. 6.3. Directional distribution of the displacement height in 1987 at the investigation site. The *thin curves* indicate the single standard deviation. The *outer circle* corresponds approximately to the canopy height

cause of the differing flow conditions (see above), the displacement height was determined as a function of the wind direction and was used accordingly in the evaluation of the deposition experiments. The distribution ascertained in 1987 is shown in Fig. 6.3 (Michaelis et al. 1988, 1989; Schönburg and Mengelkamp 1989; Mengelkamp and Schönburg 1990). The result is based on the measurement of 4395 profiles. The average annual value is 20.2 ± 4.7 m (1988: 19.9 ± 4.9 m). For wind directions from NNE to SSE, the mean amounts to 18.9 m, for the other directions it is 21.2 m. These findings suggest that the disturbing influence of the edge of the forest causes an artificial enhancement of the displacement height. The value d = 18.9 m agrees quite well with data reported by Thom et al. (1975), Garratt (1980) and Hicks (1985) who specified for the ratio of displacement height to canopy height values between 0.60 and 0.92.

If d is known, the quantities u_* and Θ_* as well as the stability functions $\Phi_{M,H}$ can be determined by means of the Eqs. (15) and (16) using an iteration procedure over L. Since in the present study three measurement points above the canopy were available, the gradients between these heights were taken into account in a weighted manner (Schönburg and Mengelkamp 1989; Mengelkamp and Schönburg 1990). Let the indices (n, m) label the pairs (1, 2), (2, 3) and (1, 3) of measured values and $S_{n,m}$ denote the corresponding gradients, then Eq. (15) may be written in the form:

$$S_{u(n,m)} = \frac{\left(\ln z_m - \Psi_{M,m}\right) - \left(\ln z_n - \Psi_{M,n}\right)}{u_m - u_n}. \tag{20}$$

The mean value over the total height follows as:

$$\langle S_u \rangle = \frac{S_{u(1,2)} + S_{u(2,3)} + 2\, S_{u(1,3)}}{4}. \tag{21}$$

An analogous procedure was applied in the case of Eq. (16). The parameters u_* and Θ_* can then be calculated from the relationships:

$$u_* = \frac{\kappa}{\langle S_u \rangle} \text{ and } \Theta_* = \frac{\kappa}{R\langle S_\Theta \rangle}. \tag{22}$$

6.4 Fluxes in the Stand and the Seepage Water

The crown compartment represents the locus for the first interaction of atmospheric pollutants with the plant organisms. In order to quantify the processes involved, the mass fluxes beneath the trees were determined in the form of the throughfall, which in the case of a spruce stand is equivalent to the canopy drip, since the stem flow within the scope of the experimental errors is negligible. When comparing the fluxes above and below the crown compartment, it has to be considered that the crown sphere accumulates atmospheric constituents on the needles or leaves also in the absence of rainfall events by the dry mechanisms of interception and sedimentation. These substances are wholly or in part transferred from the crown to the soil during the next rainfall. Thus, on the one hand, elements can be gradually accumulated in the crown sphere, whereas on the other hand elements can also be leached out from the needles or leaves, particularly under the influence of acid deposition. An independent measurement of the interception is possible with the aid of a sodium-balance approach which uses this element as an inert tracer (Ulrich 1983).

To determine the exit of substances out of the ecosystem via the seepage water, the so-called chloride method was applied which proceeds on the assumption that chloride is taken up by the plants only to a small degree and that there is no long-term interaction with the soil matrix (Brumme 1986). The flux of chloride F_{Cl} follows from the product of the water content β, the percolation velocity v and the chloride concentration c_{Cl}:

$$F_{Cl} = \beta \, v c_{Cl} . \tag{23}$$

The initial value of F_{Cl} is known from the throughfall measurement; c_{Cl} is determined in the seepage water so that the water transport βv can be calculated. Thus, with the aid of an analogous equation for the flux F_j of the element j, this flux can be derived as a function of depth from the concentrations c_j in the seepage water.

References

Brückmann A (1988) Radionuklidbilanz von vier Waldökosystemen nach dem Reaktorunfall in Tschernobyl und eine Bestimmung der trockenen Deposition. Diplomarbeit, Forstwissenschaftlicher Fachbereich, Universität Göttingen

Brumme R (1986) Modelluntersuchung zum Stofftransport und Stoffumsatz in einer Terra fusca-Rendzina auf Muschelkalk. 4. Die quantitative Ermittlung der Wasseraufnahme durch die Wurzeln mittels der Chlorid-Methode. In: Ulrich B (ed) Berichte des Forschungszentrums Waldökosysteme/Waldsterben. Universität Göttingen. Reihe A, Bd 24, pp 55–71

Businger JA, Wyngaard JC, Izumi Y, Bradley EF (1971) Flux profile relationships in the atmospheric surface layer. J Atmos Sci 28:181–189

Chamberlain AC (1966) Transport of lycopodium spores and other small particles to surfaces. Proc R Soc Lond A 296:45–70

Dolske DA, Gatz DF (1984) Field intercomparison of sulfate dry deposition monitoring and measurement methods: preliminary results. In: Hicks BB (ed) Deposition both wet and dry. Butterworth, Boston, pp 121–131

Garratt JR (1980) Surface influence upon vertical profiles in the atmospheric near-surface layer. Q J R Meteorol Soc 106:803–819

Gravenhorst G, Waraghai A (1990) Depositionsgeschwindigkeit luftgetragener Partikel für einen Fichtenbestand. VDI Berichte 837. VDI Verlag, Düsseldorf, pp 119–127

Grosch S, Schmitt G (1988) Experimental investigations on the deposition of trace elements in forest areas. In: Grefen K, Löbel J (eds) Environmental meteorology. Kluwer, Dordrecht, pp 201–216

Hertlein F (1990) Untersuchungen zur Anwendung der Gradientenmethode auf luftgetragene Partikel. Diplomarbeit, Fachbereich Angewandte Naturwisssenschaften, Fachhochschule Lübeck

Hicks BB (1985) Application of forest-atmosphere turbulent exchange information. In: Hutchinson BA, Hicks BB (eds) The forest-atmosphere interaction. Kluwer, Dordrecht, pp 631–644

Hicks BB, Wesely ML (1978) An examination of some micrometeorological methods for measuring dry deposition. US EPA-Rep, EPA-600/7-78-116

Höfken KD, Gravenhorst G (1983) Untersuchung über die Deposition atmosphärischer Spurenstoffe an Buchen- und Fichtenwald. In: UBA-Berichte 6/83, II. Schmidt-Verlag, Berlin, pp 1-212

Jonas R (1983) Ablagerungsgeschwindigkeit von Aerosolen und Gasen auf Vegetation und ebene Oberflächen. In: Arbeitsgemeinschaft der Großforschungseinrichtungen (AGF)(ed) Luftreinhaltung - Luftverschmutzung. Thenée Druck, Bonn, pp 24–26

Jonas R (1984) Ablagerung und Bindung von Luftverunreinigungen an Vegetation und anderen atmosphärischen Grenzflächen. Kernforschungsanlage Jülich, Jül-1949

Jonas R, Vogt KJ (1982) Untersuchungen zur Ermittlung der Ablagerungsgeschwindigkeit von Aerosolen auf Vegetation und anderen Probennahmeflächen. Kernforschungsanlage Jülich, Jül-1780

Mengelkamp HT, Schönburg M (1990) Vertikale turbulente Flüsse und trockene Deposition über einem Wald. Landwirtschaftliches Jahrbuch 67. Jahrg, Sonderheft, pp 59–63

Michaelis W, Schönburg M, Stößel RP (1988) Trocken- und Naßdeposition von Schwermetallen und Gasen. In: Bauch J, Michaelis W (eds) Das Forschungsprogramm Waldschäden am Standort "Postturm", Forstamt Farchau/Ratzeburg. GKSS Forschungszentrum Geesthacht, GKSS 88/E/55, pp 19–59

Michaelis W, Schönburg M, Stößel RP (1989) Deposition of atmospheric pollutants into a North German forest ecosystem. In: Georgii HW (ed) Mechanisms and effects of pollutant-transfer into forests. Kluwer, Dordrecht, pp 3–12

Oliver HR (1971) Wind profiles in and above a forest canopy. Q J R Meteorol Soc 97:548–553

Paulson CA (1970) The mathematical representation of wind speed and temperature profiles in the unstable atmospheric surface layer. J Appl Meteorol 9:857–861

Schönburg M, Mengelkamp HT (1989) Turbulente Flüsse in der atmosphärischen Grenzschicht über einem Waldgebiet. In: Arbeitsgemeinschaft der Großforschungseinrichtungen (AGF) (ed) Wechselwirkung Atmosphäre–Biosphäre. Thenée Druck KG, Bonn, pp 5–7

Sehmel GA (1980) Particulate and gas dry deposition: a review. Atmos Environ 14:983–1011

Sehmel GA, Hodgson WH (1974) Predicted dry deposition velocities. In: Energy Research and Development Administration (ed) Atmosphere-surface exchange of particulate and gaseous pollutants. NTIS CONF-740921,US Department of Commerce, Richland, Washington, pp 399–422

Tajchman SJ (1981) Comments on measuring turbulent exchange within and above forest canopy. Bull AMS 62(11):1550–1559

Thom AS, Stewart JB, Oliver HR, Gash HC (1975) Comparison of aerodynamic and energy budget estimates of fluxes over a pine forest. Q J R Meteorol Soc 101:93–105

Ulrich B (1983) Interaction of forest canopies with atmospheric constituents: SO_2, alkali and earth alkali cations and chloride. In: Ulrich B, Pankrath J (eds) Effects of accumulation of air pollutants in forest ecosystems. Reidel, Dordrecht, pp 33–45

Waraghai A, Gravenhorst G (1989) Dry deposition of atmospheric particles to an old spruce stand. In: Georgii HW (ed) Mechanisms and effects of pollutant-transfer into forests. Kluwer, Dordrecht, pp 77–86

7 Trace Elements in Rainwater: Concentrations and Wet Deposition

7.1 Concentrations of Dissolved and Particulate Constituents

When studying the impact of atmospheric pollutants on terrestrial ecosystems, not only anthropogenic, but also geogenic constituents should be taken into account, since in the presence of harmful substances, for instance acid precipitation, the balance of conducive elements such as nutrients may be impaired. Knowledge of their deposition is therefore also of interest. This will be discussed in detail in Chapter 11. Consequently, a broad spectrum of elements was recorded in this study with regard to both wet and dry deposition. It is also useful to separately determine the dissolved and the particulate phases in rainwater.

In Table 7.1 the trace analytical results for two weekly rainwater samples are summarized. All data are given in ng/g. As can be seen, a comprehensive spectrum of elements could be determined with the methods applied. The two samples were selected in consideration of the following criteria: same season, but with a time interval of a few years (36th calendar week 1987; 35th calendar week 1990); similar amount of precipitation (28.5 mm and 19.4 mm, respectively); no or negligible rainfall in the preceding week and approximately equal wind directions during the rainfall. In spite of these restrictions, there are distinct differences with regard to both the concentration levels and the partition of the two phases, even in the case of geogenic elements. This demonstrates that the composition of the samples is a rather complex outcome of the meteorological conditions, the source-receptor relationships and atmospheric chemistry. For a few anthropogenic elements, in particular sulphur and lead, the analytical results of Table 7.1 already suggest a decrease in the atmospheric pollution. In view of the above conclusions, however, the derivation of trends should be reserved to the analysis of properly chosen long-term means (cf. Sect. 7.2) in which short-term variations are averaged out. Single measurements do not reveal conclusive evidence.

Figure 7.1 illustrates the courses of the concentrations during a four-week rainfall event in 1987 for a few selected elements. The meteorological conditions were as follows:

	Precipitation	Prevailing wind direction
Preceding week	3.3 mm	230°–290°
36th week	28.5 mm	200°–250°
37th week	26.2 mm	180°–235°
38th week	29.2 mm	110°–235°
39th week	23.6 mm	180°–240°

Clearly discernible is a decline in the concentrations after the beginning of the event as a result of washout effects in the atmosphere. This phenomenon is even more pronounced if samples are taken over short time intervals in the first hours of the rainfall period (Nürnberg et al. 1982).

Sulphur in the dissolved phase has to be ascribed mainly to the sulphate content. This follows from the diagram shown in Fig. 7.2, where the results of the trace element analyses are plotted against the sulphur concentrations derived using ion chromatography (cf. Chap. 8). The regression lines obtained for the data from 1988 to 1992 deviate from the angular bisector by a

Table 7.1. Typical concentrations of trace elements in weekly rainwater samples. I: 36th calendar week 1987, 28.5 mm precipitation; II: 35th calendar week 1990, 19.4 mm precipitation. All data in ng/g; d.l. = detection limit; n.d. = not determined

	I			II		
Element	Dissolved phase	Particulate phase	Total	Dissolved phase	Particulate phase	Total
Na	580	n.d.	–	1290	n.d.	–
Mg	170	n.d.	–	230	n.d	–
Al	130	n.d	–	n.d.	n.d.	–
S	2002	13.8	2016	1229	7.9	1237
K	95.2	36.5	132	104	16.2	120
Ca	402	12.1	414	329	7.7	337
Ti	< d.l.	9.2	9.2	0.26	5.0	5.3
V	0.79	0.34	1.1	0.75	0.10	0.85
Cr	0.36	0.43	0.79	0.71	0.18	0.89
Mn	4.8	1.7	6.5	4.2	0.45	4.7
Fe	31.9	142	174	9.8	31.1	41
Ni	0.60	0.20	0.80	< d.l.	0.13	0.13
Cu	2.6	0.57	3.2	1.2	0.27	1.5
Zn	26.3	1.57	27.9	21.4	0.43	21.8
As	0.51	0.07	0.58	0.01	0.01	0.02
Se	0.36	0.01	0.37	0.21	0.01	0.22
Rb	0.28	0.23	0.51	0.09	0.09	0.18
Sr	2.1	0.29	2.4	1.5	0.12	1.6
Y	< d.l.	0.06	0.06	0.11	0.02	0.13
Zr	< d.l.	0.99	0.99	0.78	0.14	0.92
Nb	< d.l.	0.04	0.04	< d.l.	0.01	0.01
Mo	0.13	0.05	0.18	< d.l.	0.02	0.02
Cd	0.19	< d.l.	0.19	0.28	< d.l.	0.28
Sn	0.60	0.19	0.79	< d.l.	< d.l.	<d.l.
Sb	< d.l.	0.07	0.07	0.34	< d.l.	0.34
Ba	2.5	0.84	3.3	1.5	0.47	2.0
Pb	9.3	1.64	10.9	2.17	0.34	2.5

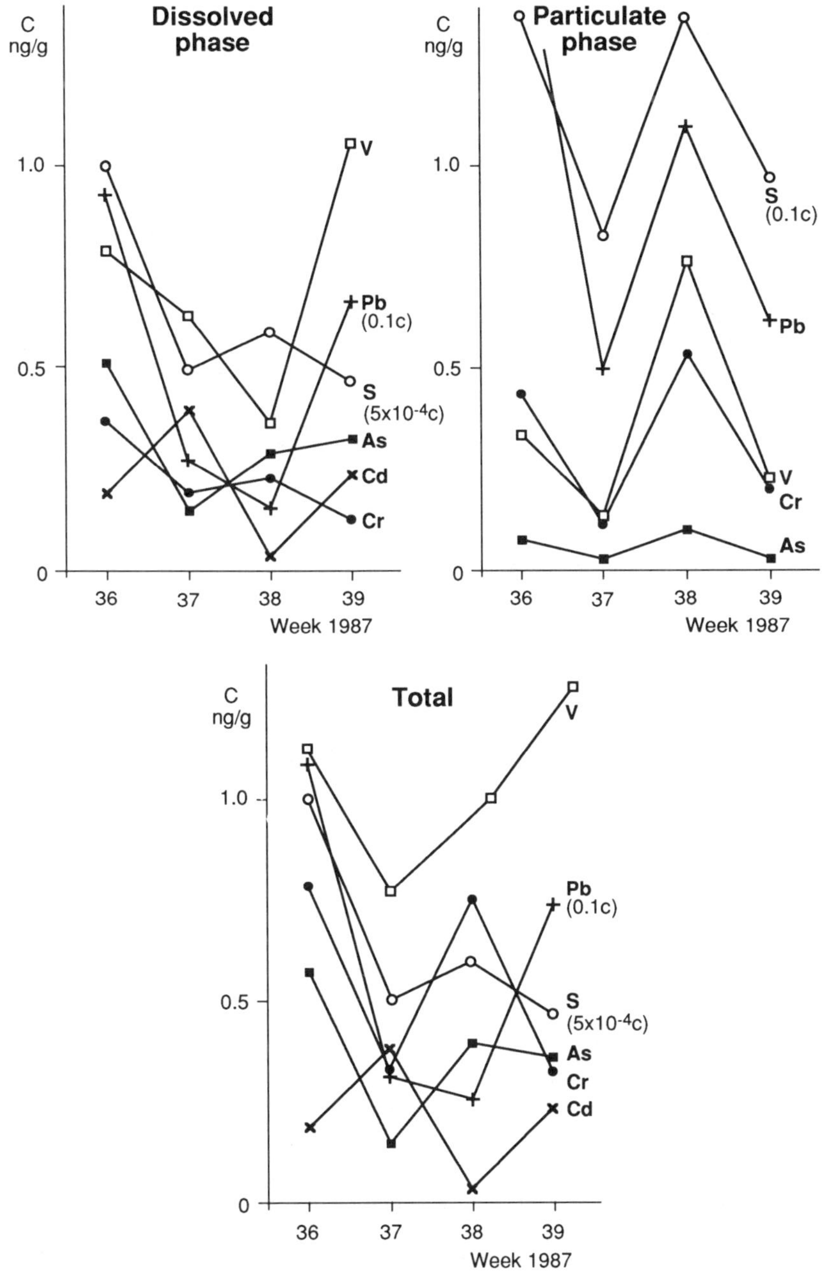

Fig. 7.1. Variations of the concentrations of a few selected elements during a 4-week rainfall event in 1987

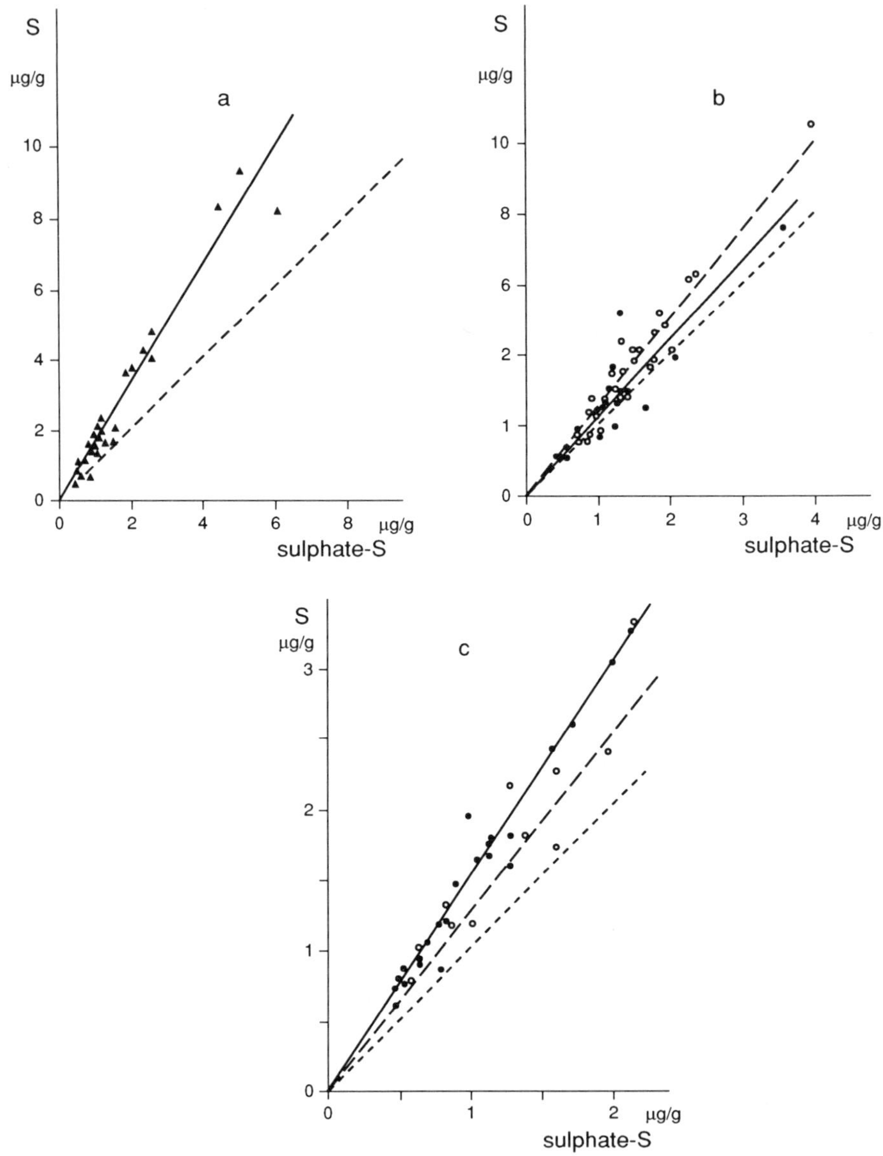

Fig. 7.2. Correlations between the sulphur concentrations as obtained from trace element analysis and sulphate ion chromatography, respectively. *Dotted lines* Bisectors. Note the change of the scale. **a** Data from 1988. **b** *Open circles* and *dashed line* data from 1989; *full circles* and *full line* data from 1990. **c** *Full circles* and *full line* data from 1991; *open circles* and *dashed line* data from 1992

percentage ranging from 10 to 40 % of the total concentration. Possible constituents other than sulphate could be due to sulphur in the oxidation state +4 (S IV) (Jaeschke 1987). Under this notation the components physically dissolved SO_2, hydrogen sulphite HSO_3^- and sulphite SO_3^{2-} are summarized. Another potential constituent may be the complex compound hydroxymethane sulphonate $HOCH_2SO_3^-$ which is able to inhibit the liquid phase oxidation of SO_2 to SO_4^{2-} (Munger et al. 1983; Beltz et al. 1986; Lammel and Metzig 1989). The concentration of natural sulphur compounds is low compared to man-made sulphur (Georgii 1978; Seinfeld 1980). Figure 7.2 also gives a clear indication of a decrease in the sulphur concentration during the investigaton period. This trend will be confirmed by the results presented in Section 7.2.

The pollutant concentration shows a distinct dependence on the wind direction. This is exemplified in Fig. 7.3 for the elements sulphur and lead. Only those weekly samples which could be related to rather stable meteorological conditions were taken into account. Therefore, the plots have to be understood in a more qualitative than quantitative manner. It is evident, however, that sources located southwest to northwest of the measuring station determine to a high degree the immission conditions. If the distribution of the amount of precipitation as a function of the wind direction is also taken into account (cf. Fig. 7.5), it is expected that the wet deposition at the investigation site is strongly affected by the conurbation of Hamburg (cf. Fig. 2.1). This will be confirmed in the next section. Additional evidence is provided by the results of soil analyses which were performed at different distances from a large non-ferrous metallurgical plant located in the city of Hamburg (Michaelis 1986). Depth profiles of a few elements as obtained close to the plant and 30 km east of it are shown in Fig. 7.4. Significant effects appear in the case of Zn, As and Sb. Near the plant the threshold values are clearly exceeded for the latter two elements.

Figure 7.3 and analogous data for other measuring periods also indicate high concentrations in the case of rainfall events with wind around southeast, in particular for sulphur. These findings have to be attributed to high pollutant emissions in the southern brown-coal mining and industrial area of the former German Democratic Republic (cf. Chaps. 9 and 12). After the German unification the concentrations exhibited a decreasing tendency. It should be pointed out, however, that even in the early years of the present investigation these rainfall events had only little influence on the overall wet deposition, since rainfall with wind from the southeast is rather rare and the amount of precipitation comparatively low.

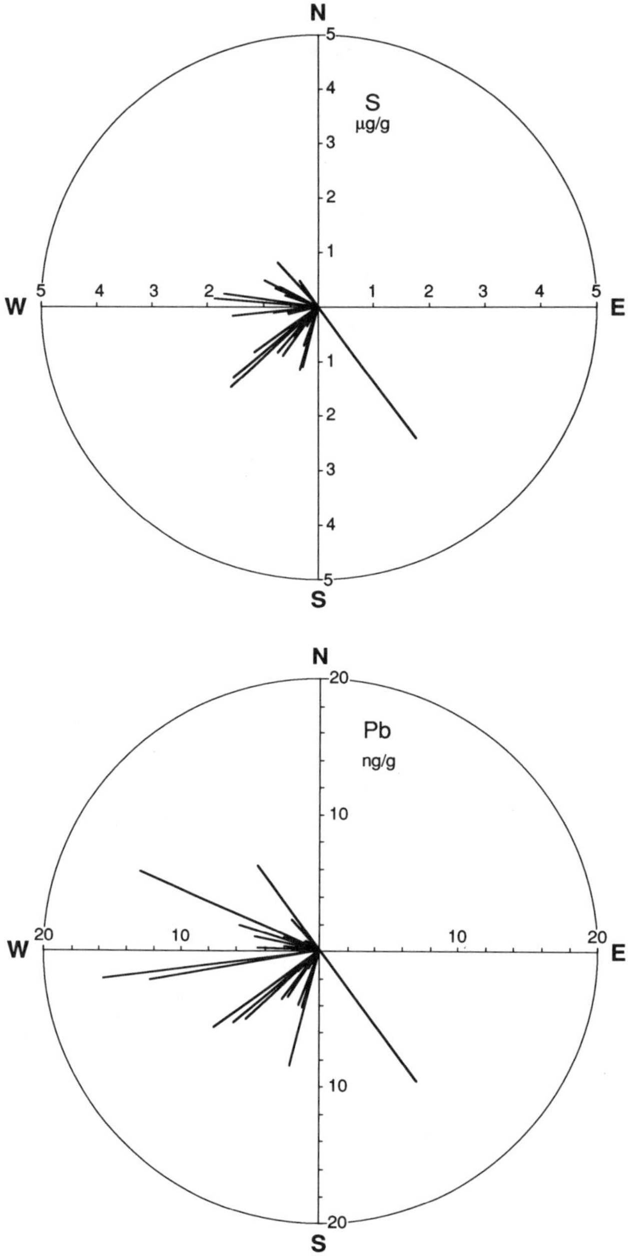

Fig. 7.3. Concentrations of sulphur (µg/g) and lead (ng/g) in the liquid phase of rainwater vs. wind direction. Data from 1988

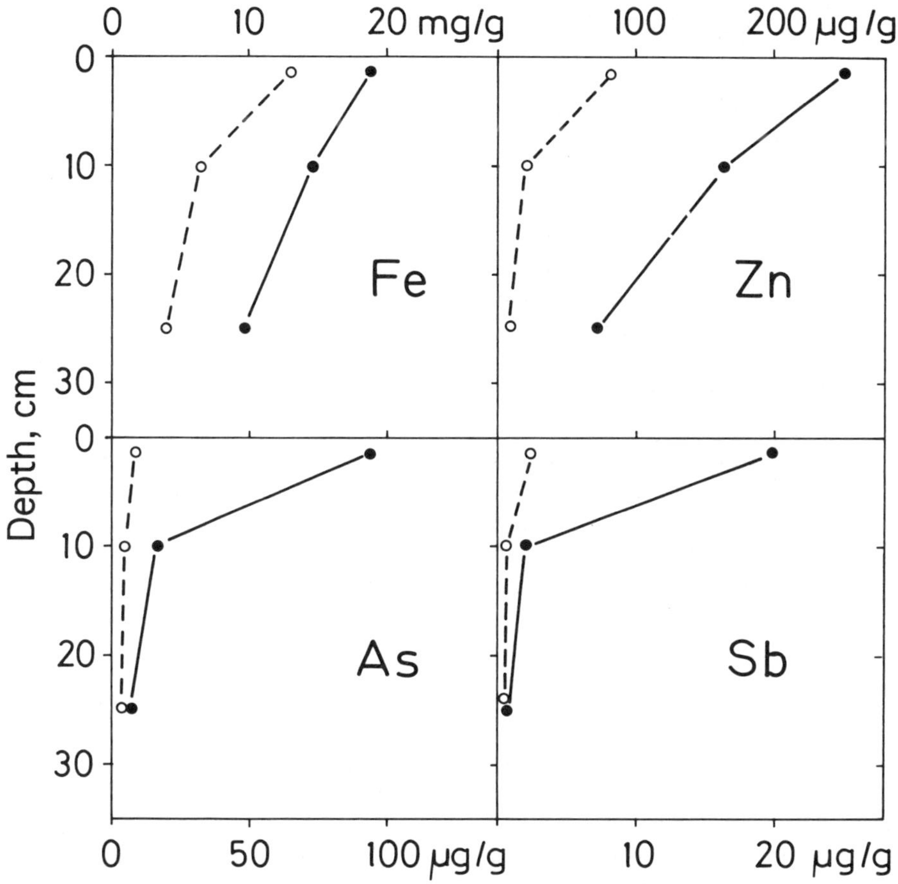

Fig. 7.4. Depth profiles of iron, zinc, arsenic and antimony in soil close to a non-ferrous metallurgical plant (*full curve*) and 30 km east of it (*dashed curve*). All values refer to the dry substance

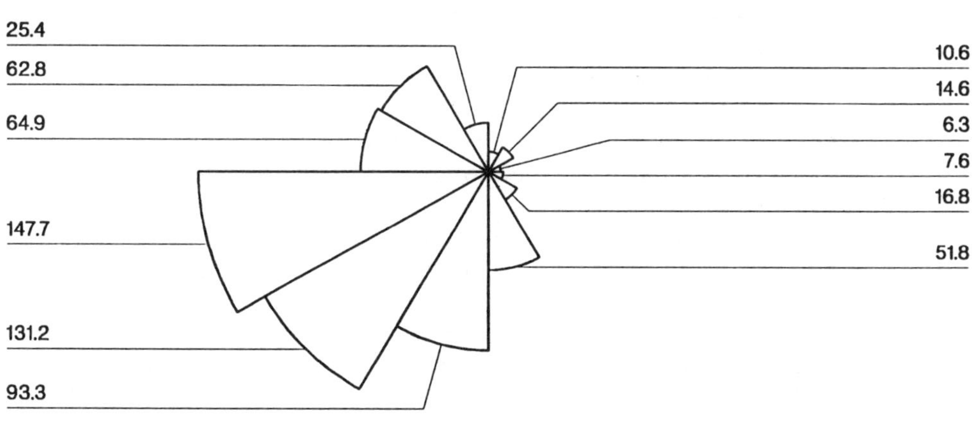

Fig. 7.5. Distribution of the amount of precipitation in mm/year as a function of the wind direction. Data from 1991 collected in sectors of 30°

7.2 Wet Deposition

7.2.1 Meteorological Aspects

The distribution of the amount of precipitation as a function of the wind direction is exemplified in Fig. 7.5 for the year 1991. The pronounced anisotropy gives rise to the fact that the wet deposition is predominantly influenced by emission sources located in a sector between 210° and 270°. This, in particular, includes the conurbation of the city of Hamburg (cf. Fig. 2.1). Typical plots of the deposition of sulphur and lead are presented in Fig. 7.6. The data refer to the total deposition, i.e. the sum of the dissolved and particulate phases. A comparison with Fig. 7.3 clearly illustrates the influence of the rainfall distribution shown in Fig. 7.5.

The amount of precipitation also exhibits a distinct seasonal variation with maxima mostly in the second or third quarter. Data measured at the nearby town of Ratzeburg (cf. Fig. 2.1) for the period 1987 to 1991 are presented in Fig. 7.7 (Deutscher Wetterdienst, Wetteramt Schleswig). Together with fluctuations in the emissions these variations determine to a high degree the seasonal behaviour of the wet deposition (cf. Sect. 7.2.3).

7.2.2 Long-Term Mean Values

The results obtained by TXRF for the long-term means of the daily wet deposition of trace elements are summarized in Table 7.2 (Michaelis et al. 1992). The data cover the period April 1987 to March 1991. Dissolved and particulate phases are specified separately. Long-term means have the advantage that short-time fluctuations are eliminated so that a rather representative overview is obtained. On the other hand, they hide temporal trends, a knowledge of which is of great importance for environmental policy. For the latter purpose, the annual means observed over a longer period are most suitable (cf. Sect. 7.2.4).

For examining the consistency of the data it is useful to compare the findings with results obtained at other North German investigation sites. These are presented in Table 7.3 where measurements from the "Postturm" project for the period April 1987 to January 1990 are listed together with data from Pellworm Island in the German Bight, from the town of Schleswig in the northern part of Schleswig-Holstein, from List on the North Frisian island of Sylt and finally from the city of Hamburg. The investigation on Pellworm Island was performed in 1984/1985 by the GKSS Research Centre using equipment and methods similar to those of the present study (Stößel et al. 1985; Michaelis and Stößel 1986; Stößel 1987; Michaelis et al. 1988, 1990). The goal of the measurements on this island was to estimate the atmospheric heavy metal transport into the German Bight. The data cover the period May 1984 to July 1985. Pellworm Island is situated about 165 km

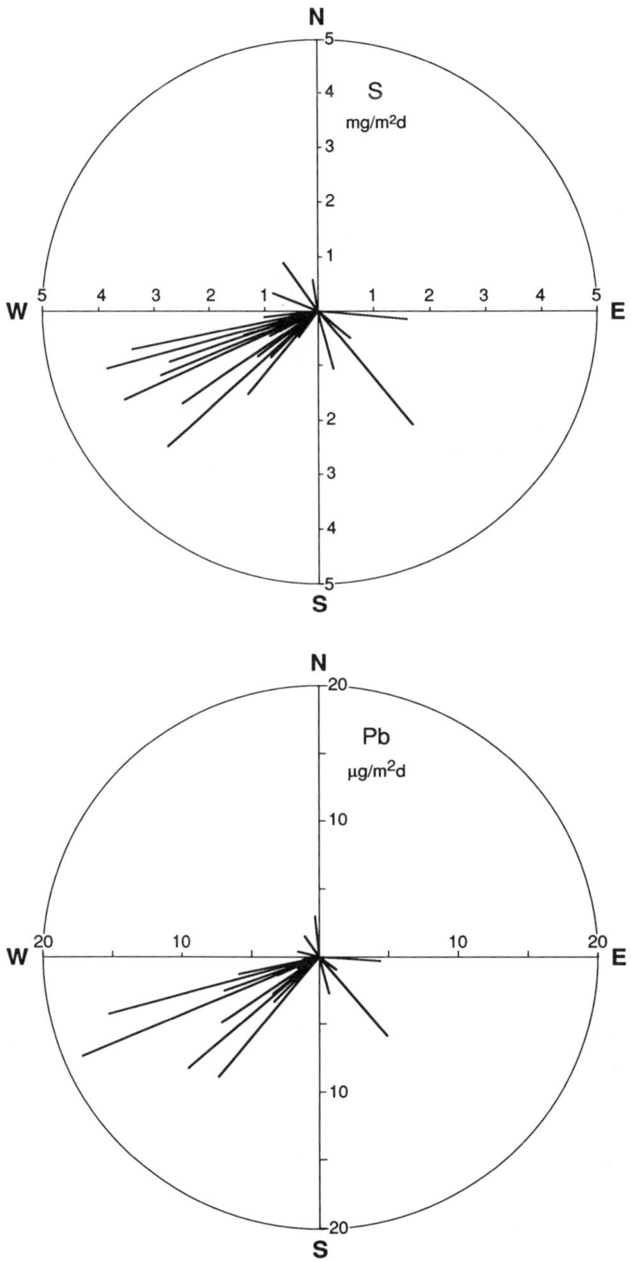

Fig. 7.6. Total wet deposition of sulphur (mg/m²d) and lead (µg/m²d) vs. wind direction. Data from November 1990 to November 1991

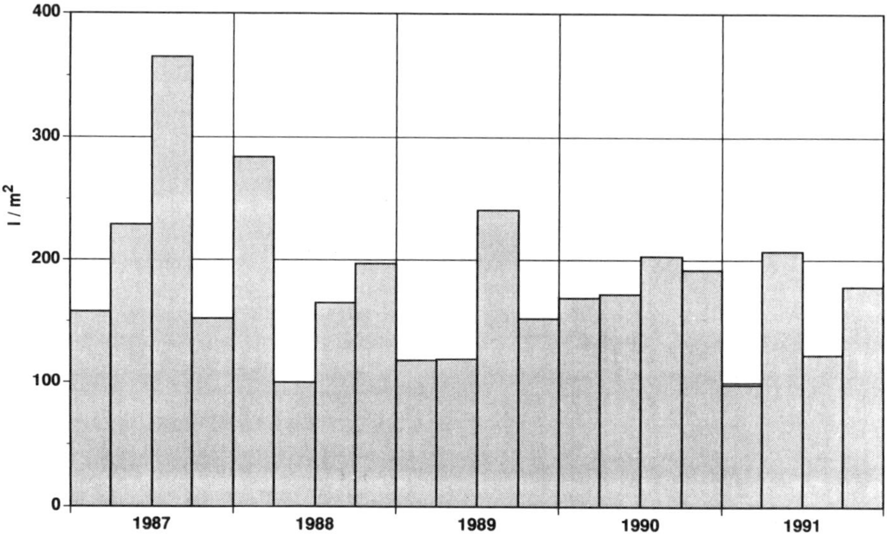

Fig. 7.7. Variation in the amount of precipitation during the period 1987 to 1991

Table 7.2. Long-term mean values of the daily wet deposition of trace elements in $\mu g/m^2d$. Measuring period April 1987 to March 1991

Element	Dissolved phase	Particulate phase	Total
S	2370	40	2410
K	170	43	210
Ca	630	34	660
Ti	0.64	12.6	13
V	1.4	0.4	1.8
Cr	0.53	0.83	1.3
Mn	8.5	2.1	11
Fe	33	147	180
Ni	1.3	0.51	1.8
Cu	3.6	0.82	4.4
Zn	50	2.8	53
As	1.0	0.10	1.1
Se	0.47	0.02	0.5
Rb	0.50	0.29	0.8
Sr	4.2	0.52	4.7
Y	0.04	0.074	0.11
Zr	0.15	0.85	1.0
Nb	0	0.050	0.05
Mo	0.16	0.06	0.22
Cd	0.55	0.04	0.60
Sn	0.44	0.21	0.65
Sb	0.04	0.023	0.06
Ba	6.0	1.7	7.7
Pb	9.3	1.7	11

Table 7.3. Intercomparison of daily wet deposition long-term means for some trace elements as measured at several North German sites (see text). All data in µg/m²d

Element	"Postturm"	Pellworm	Schleswig		List/Sylt	Hamburg	
	Total deposition		Dissolved phase only				
	April 1987– January 1990	May 1984– July 1985	1980	1981	1981	1980	1981
S	2790	4040	–	–	–	–	–
K	219	383	–	–	–	–	–
Ca	725	730	–	–	–	–	–
V	2.0	2.7	–	–	–	–	–
Cr	1.1	0.47	–	–	–	–	–
Mn	12	7.5	–	–	–	–	–
Fe	208	163	–	–	–	–	–
Ni	1.9	2.0	–	–	–	–	–
Cu	4.1	2.5	6.1	–	–	19.3	–
Zn	72	27.6	46.6	–	–	80.9	–
As	1.3	0.81	–	–	–	–	–
Se	0.54	0.45	0.07[a]	–	–	0.16[a]	–
Cd	0.65	0.50	0.6	0.7	0.5	1.0	0.7
Pb	13	10.3	30	36	18	48	38

[a] Se (IV) only.

Table 7.4. Comparison of the total daily wet deposition (cf. Table 7.3) of some elements with available literature data (dissolved phase only) reported for southwest German forest ecosystems (see text). All data in µg/m[a] d

Element	"Postturm"	Schönbuch		Rotenfels		Freudenstadt	
	1987/1990	1984/1985	1985/1986	1984/1985	1985/1986	1984/1985	1985/1986
Na	1200	260	990	718	1570	–	1360
Mg	184	203	252	282	320	290	466
K	219	580	1190	680	784	1150	3120
Ca	725	562	1200	943	1250	578	1380
Mn	12	29	58	27	47	40	196
Zn	72	38	99	467	517	133	72
Cd	0.65	–	–	0.60	0.82	0.90	0.63
Pb	13	–	–	28	31	20	16

northwest of the Postturm investigation site. The other measurements listed in Table 7.3 were carried out by the KFA Research Centre Jülich (Nürnberg et al. 1982, 1983) using the same sampling technique, but with voltammetric methods for trace analysis (Nürnberg 1982). In these studies only the dissolved phase was investigated and the annual means were listed in Table 7.3. The measuring site Schleswig lies about 110 km north-northwest, List on the island of Sylt about 210 km northwest of the Postturm station. When comparing the data in Table 7.3, a substantial consistency can be discerned, if the different geographic positions and distances to conurbations, the phases considered, the date of the measurements and the temporal trends for several pollutants (cf. Sect. 7.2.4) are taken into account. It may be useful to point out in this connection that the Institute of Physics of the GKSS Research Centre also performed extensive model calculations on the atmospheric transport of trace substances and their deposition over Europe (Petersen et al. 1989, 1995; Petersen and Krüger 1993).

It is also of particular interest to compare the Postturm data with results obtained by other research groups for forest ecosystems in southwest Germany. Adam et al. (1987) have summarized the findings for several ions (cf. Chap. 8) and trace elements at a total of 11 measuring sites and for a few hydrological years. The trace element data from three of these sites are compared in Table 7.4 with results of the present study. Measuring periods closest to the beginning of the Postturm project have been selected. Schönbuch is a dense woodland situated between the towns of Stuttgart and Tübingen. It is considered to be a little polluted area and exhibits only moderate forest decline symptoms. The measuring site Rotenfels lies in the northwestern region of the Black Forest with access to air masses from the Rhine plain. In places, the impact on the forest is severe with needle losses of more than 50 %. Marked deficiencies in magnesium and manganese as well as a moderate deficiency in calcium were detected in the needles. Freudenstadt is situated beyond the main ridge of the Black Forest in an apparently sheltered position. Nevertheless, severe forest decline has been ascertained. Analyses of needles in this case also revealed a shortage of magnesium and calcium.

On the whole Table 7.4 suggests a more favourable supply of nutritive elements in the southwestern forest ecosystems, while the pollutant input seems to be of the same order of magnitude or even higher in spite of possible temporal trends. Most probably the extent of proton deposition, washout effects in the crown compartment, element mobilization in the soil and discharge from the ecosystem via the seepage water control to a high degree the element budget (cf. Chaps. 8, 11). Moreover, the dry deposition of trace elements has to be included in these considerations (cf. Chaps. 9, 10).

7.2.3 Seasonal Variations

The wet deposition shows a marked seasonal variation (Michaelis et al. 1992; Pepelnik et al. 1993). This is demonstrated in Fig. 7.8 for the total period of the present study using the elements S, As, Cd and Pb as examples. The data are plotted quarterly, and the dissolved and particulate phases are specified separately. A comparison with Fig. 7.7 demonstrates the strong influence of the respective amount of precipitation.Maxima occur in the second or third quarter. Figure 7.8 also reveals a trend of decreasing atmospheric pollution for these elements. This will be quantified in a more evident manner in the following section.

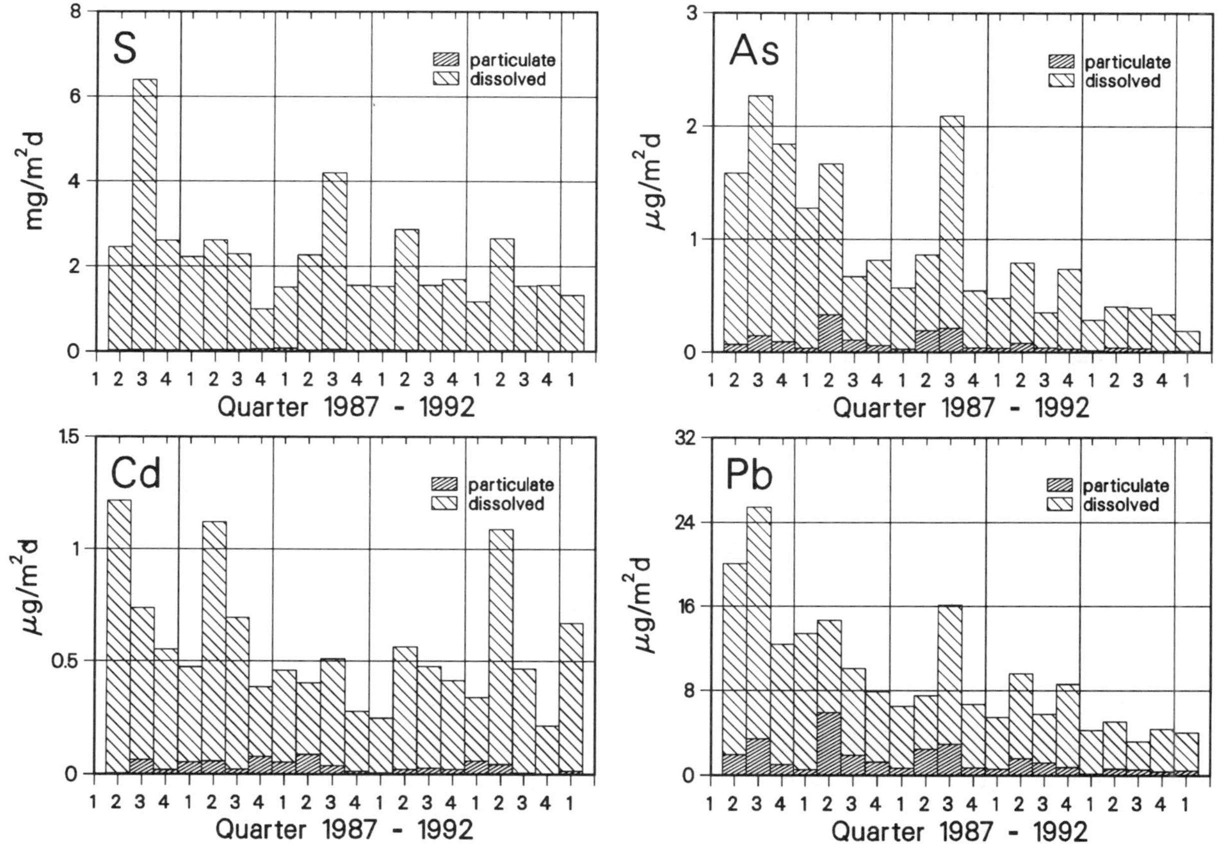

Fig. 7.8. Seasonal variation in the daily wet deposition (dissolved and particulate phase) of S, As, Cd and Pb during the investigation period

7.2.4 Annual Means and Long-Term Trends

The annual means of the total daily wet deposition during the investigation period are summarized in Table 7.5 for the majority of the elements determined. It is shown that several anthropogenic pollutants feature a clear decreasing tendency. This is particularly evident for sulphur, arsenic, cadmium and lead. The sulphur deposition significantly decreased by more than 50 % between 1987 and 1992 as a result of environmental protection measures with regard to the reduction in sulphur emissions, in particular, from large-scale fuel combustion facilities. There is no doubt that sulphur is important to the metabolism of plants. It is a structural element of proteins and enzymes; and sulphate, which makes up the main sulphurous constituent of the rainwater (cf. Sect. 7.1), can be taken up by the plants via the roots. On the other hand, however, sulphate is a reaction product of sulphur dioxide which is released, for instance, by combustion processes, and this gas – together with nitrogen oxides – is substantially responsible for the acid rain (cf. Chap. 8) and also has negative effects on the carbon dioxide budget (cf. Chap. 13). A similarly pronounced decreasing tendency has been observed for the heavy metal lead. The wet deposition of this element declined by

Table 7.5. Annual mean values of the total daily wet deposition of trace elements. All data in µg/m²d

Element	1987[a]	1988	1989	1990	1991	1992[b]
S	3830	2001	2440	1940	1775	1485
K	248	169	247	231	162	245
Ca	693	604	876	555	404	460
Ti	11.4	14.3	19.7	9.6	9.4	9.3
V	2.3	1.9	1.7	1.6	1.4	1.6
Cr	1.2	0.81	1.3	2.0	2.3	2.6
Mn	13.8	9.8	11.7	9.5	8.2	4.9
Fe	221	201	203	98	209	83
Ni	3.4	0.75	1.8	1.9	1.6	1.5
Cu	5.8	3.2	3.8	6.3	2.5	3.3
Zn	89	65	67	15	62	37
As	1.9	1.1	1.1	0.6	0.5	0.32
Se	0.7	0.5	0.5	0.4	0.3	0.3
Rb	0.9	0.8	0.9	0.6	0.6	0.4
Sr	5.6	4.7	5.2	3.4	3.0	2.2
Y	0.08	0.21	0.12	0.05	0.08	0.03
Zr	0.8	1.2	1.2	0.8	0.5	0.4
Nb	0.05	0.06	0.08	0.02	0.02	0.03
Mo	0.35	0.21	0.16	0.16	0.18	0.08
Cd	0.84	0.65	0.41	0.44	0.53	0.69
Sn	1.1	0.35	0.95	0.18	0.70	0.25
Sb	0.12	0.04	0.05	0.05	0.09	0.02
Ba	12	7.0	7.3	4.6	4.4	4.7
Pb	19	11	9.5	7.4	4.3	5.0

[a] April to December only
[b] January to April only

about a factor of 4 during the investigation period. The main cause of this favourable development lies in the enlarged employment of unleaded petrol in motor traffic.

In this connection, it is useful to remember that the wet deposition at the investigation site is to a large extent determined by the conurbation of the city of Hamburg (cf. Sects. 7.1, 7.2.1). This also becomes obvious for the elements arsenic and cadmium. In the middle of the 1980s, the Freie and Hansestadt Hamburg decided to implement emission-lowering measures in the industrial domain with the aim of reducing the emissions of As in the period 1987 to 1995 from 5.5 to 3.6 tons per year and those of Cd from 0.9 to 0.6 tons per year (Umweltbehörde der Freien und Hansestadt Hamburg 1990). The trends presented in Table 7.5 substantiate the success of these environmental policy measures. Geogenic elements exhibit the usual fluctuations as expected. Results of an official monitoring programme in Schleswig-Holstein (Landesamt für Wasserhaushalt und Küsten Schleswig-Holstein 1995) are in good agreement with the data obtained in the present comprehensive study.

References

Adam K, Evers FH, Littek T (1987) Ergebnisse niederschlagsanalytischer Untersuchungen in südwestdeutschen Waldökosystemen 1981–1986. In: Projekt Europäisches Forschungszentrum für Maßnahmen zur Luftreinhaltung (PEF) im Kernforschungszentrum Karlsruhe (ed). KfK-PEF 24

Beltz N, Enderle KH, Jaeschke W, Obenland H (1986) Measurements of sulfur species during the 1986 field experiment in S. Pietro Capofiume. In: Fuzzi S (ed) Heterogeneous chemistry project. Istituto Fisbat, Bologna, 1986

Georgii HW (1978) Large scale spatial and temporal distribution of sulfur compounds. Atmos Environ 12:681–690

Jaeschke W (1987) Physikalische Chemie des Niederschlags. In: Jaenicke R (ed) Atmosphärische Spurenstoffe. VCH Verlagsgesellschaft mbH, Weinheim, pp 31–76

Lammel G, Metzig G (1989) Die Säurebildung in Nebel- und Wolkenwasser. In: Arbeitsgemeinschaft der Großforschungseinrichtungen (AGF)(ed) Wechselwirkung Atmosphäre-Biosphäre. Thenée Druck, Bonn, pp 17–20

Landesamt für Wasserhaushalt und Küsten Schleswig-Holstein (ed)(1995) Ein Jahrzehnt Beobachtung der Niederschlagsbeschaffenheit in Schleswig-Holstein 1985–1994. Published by the editor, Kiel

Michaelis W (1986) Naß- und Trockendeposition von Schwermetallen. In: Technische Mitteilungen, Haus der Technik, Essen (ed) 79, 5/6:266–271

Michaelis W, Stößel RP (1986) Untersuchungen zur Schwermetalldeposition auf der Insel Pellworm – ein Beitrag zur Ermittlung des atmosphärischen Schadstoffeintrags in die Nordsee. In: GKSS Forschungszentrum Geesthacht, GKSS Jahresbericht 1986, pp 8–23

Michaelis W, Schönburg M, Stößel RP (1988) Trocken- und Naßdeposition von Schwermetallen und Gasen. In: Bauch J, Michaelis W (eds) Das Forschungsprogramm Waldschäden am Standort "Postturm", Forstamt Farchau/Ratzeburg. GKSS Forschungszentrum Geesthacht, GKSS 88/E/55, pp 19–59

Michaelis W, Pepelnik R, Rademacher P, Riebesell M (1990) Wechselwirkung zwischen Luftschadstoffen und Vegetation. In: GKSS Forschungszentrum Geesthacht, GKSS Jahresbericht 1990, pp 42–55

Michaelis W, Pepelnik R, Theopold F, Rademacher P (1992) Deposition atmosphärischer Spurenstoffe und Stoffflüsse im Ökosystem Wald. In: Michaelis W, Bauch J (eds) Luftverunreinigungen und Waldschäden am Standort "Postturm", Forstamt Farchau/Ratzeburg. GKSS Forschungszentrum Geesthacht, GKSS 92/E/100, pp 11–59

Munger JW, Jacob DJ, Waldman JM, Hoffmann MR (1983) Fogwater chemistry in an urban atmosphere. J Geophys Res 88:5109–5121

Nürnberg HW (1982) Voltammetric trace analysis in ecological chemistry of toxic metals. Pure Appl Chem 54(4):853–878

Nürnberg HW, Valenta P, Nguyen VD (1982) Wet deposition of toxic metals from the atmosphere in the Federal Republic of Germany. In: Georgii HW, Pankrath J (eds) Deposition of atmospheric pollutants. Reidel, Dordrecht, pp 143–157

Nürnberg HW, Nguyen VD, Valenta P (1983) Deposition von Säure und toxischen Schwermetallen mit den Niederschlägen in der Bundesrepublik Deutschland. Kernforschungsanlage Jülich, KFA Jahresbericht 1982/83, pp 41–53

Pepelnik R, Erbslöh B, Michaelis W, Prange A (1993) Determination of trace element deposition into a forest ecosystem using total-reflection X-ray fluorescence. Spectrochim Acta 48B(2):223–229

Petersen G, Krüger O (1993) Untersuchung und Bewertung des Schadstoffeintrags über die Atmosphäre im Rahmen von PARCOM (Nordsee) und HELCOM (Ostsee) – Teilvorhaben: Modellierung des großräumigen Transports von Spurenmetallen. GKSS Forschungszentrum Geesthacht, GKSS 93/E/28

Petersen G, Weber H, Graßl H (1989) Modelling the atmospheric transport of trace metals from Europe to the North Sea and the Baltic Sea. In: Pacyna JM, Ottar B (eds) Control and fate of atmospheric trace metals. Kluwer, Dordrecht, pp 57–83

Petersen G, Iverfeldt A, Munthe J (1995) Atmospheric mercury species over central and northern Europe. Model calculations and comparison with observations from the nordic air and precipitation network for 1987 and 1988. Atmos Environ 29(1):47–67

Seinfeld JH (1980) Lectures in atmospheric chemistry, no 12, vol 76. American Institute of Chemical Engineers, New York

Stößel RP, Michaelis W, Prange A (1985) Studies on the atmospheric heavy metal transport into the German Bight. In: CEP Consultants Ltd (ed) Int Conf on Heavy metals in the environment, Sept 10–13, 1985, in Athens, Greece, vol I. Edinburgh 1985, pp 203–205

Stößel RP (1987) Untersuchungen zur Naß- und Trockendeposition von Schwermetallen auf der Insel Pellworm. Thesis, University of Hamburg, GKSS 87/E/34

Umweltbehörde der Freien und Hansestadt Hamburg (1990)(ed) Luftreinhaltung in Hamburg – Sachstandsbericht 1990

8 Deposition of Anions and Cations Via Precipitation

8.1 Meteorology and Source-Receptor Relationship

In 1987, the average ion concentrations amounted to 6.1 µg/g for sulphate, 5.3 µg/g for nitrate, 1.05 µg/g for chloride, 0.03 µg/g for fluoride and 2.1 µg/g for ammonium. While the sulphate values decreased to 2.6 µg/g in 1991, no significant trends were observed for the other ions. The pronounced dependence of the amount of precipitation on the wind direction (cf. Fig. 7.5) suggests that the wet deposition is predominantly influenced by sources located between west and south-southwest. This is confirmed in Figs. 8.1 and 8.2 for the most important ions SO_4^{2-}, NO_3^-, Cl^- and NH_4^+. As is to be expected on the basis of Fig. 7.2, sulphate shows a behaviour similar to that of sulphur as determined by trace element analysis of the rainwater (Fig. 7.6). The contribution of the particulate phase is negligible in the case of sulphur (see Table 7.1). The sources for SO_4^{2-} lie in the western parts of Germany with a certainly marked component due to the conurbation of Hamburg. The main emitters of SO_2 are the power and district heating stations. A similar transport situation is valid for NO_3^-. Here, the main source is the NO_x-emitting motor traffic.

In contrast to the above man-made pollutants, chloride in rainwater at the investigation site is predominantly attributed to natural sea spray. The differing distribution of the wet deposition as a function of the wind direction (Fig. 8.2) and the intensification during specific events seem to support this assumption. A spatial comparison, however, may suggest the existence of anthropogenic sources as well (cf. Table 8.2). The data of the present study do not allow definite conclusions in this respect. Chloride influences the water balance of plants, and it is important for the cation-anion equilibrium. Too high an input, however, can lead to a soil salinization which is harmful to many tree species.

The main sources of ammonium are the intensive agriculture and animal keeping as well as the fertilizer and chemical industries. Ammonium is a constituent of frequently used fertilizers. In the soil it is converted by microorganisms to nitrite and finally to nitrate. This process entails additional acidification. Though the "Postturm" site is largely surrounded by agricul-

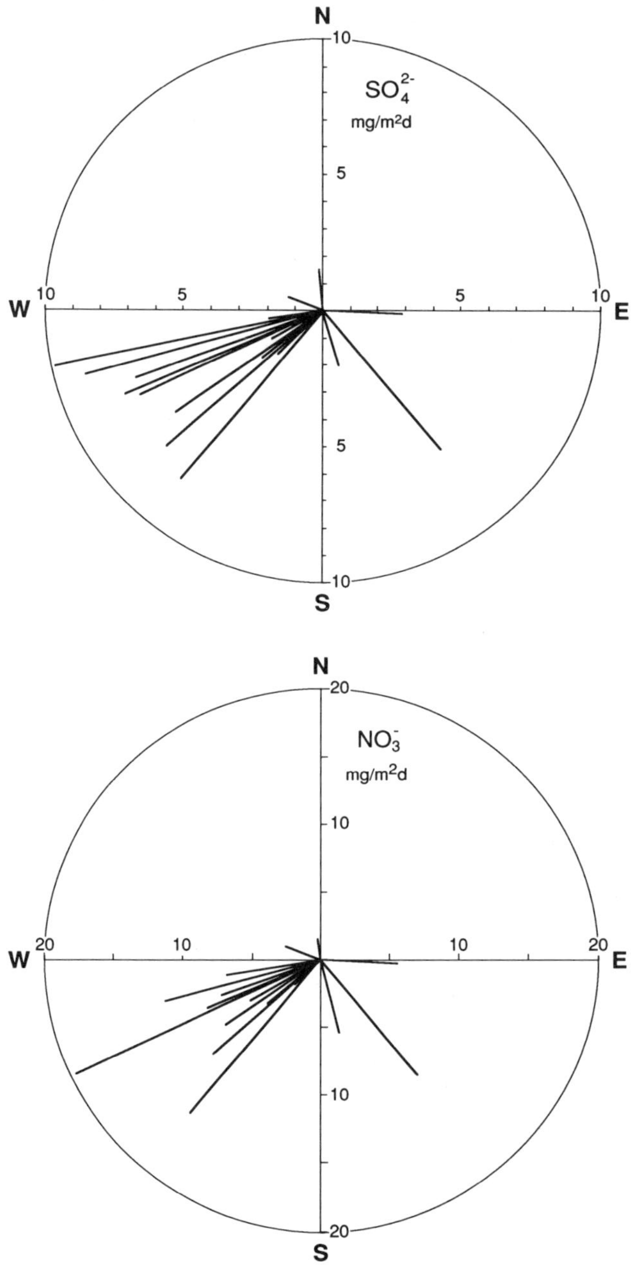

Fig. 8.1. Deposition of sulphate and nitrate (mg/m²d) vs. wind direction. Data from November 1990 to November 1991

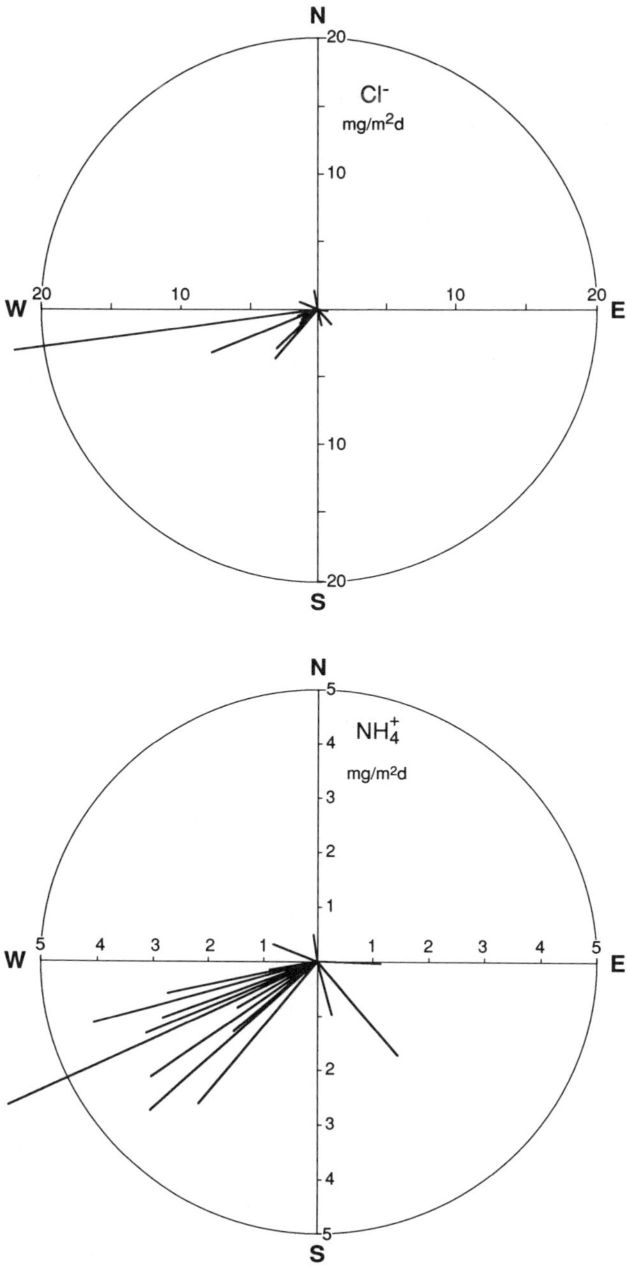

Fig. 8.2. Deposition of chloride and ammonium (mg/m²d) vs. wind direction. Data from November 1990 to November 1991

Table 8.1. Annual mean values of the daily deposition of SO_4^{2-}, NO_3^-, Cl^-, F^- and NH_4^+ in mg/m^2d

	1987[a]	1988	1989	1990	1991	1992[b]
SO_4^{2-}	11.5	5.1	5.3	5.3	3.7	3.2
NO_3^-	9.0	5.7	6.8	7.5	6.6	7.3
Cl^-	2.3	2.4	4.0	2.2	2.0	2.1
F^-	0.044	0.027	0.043	–	–	–
NH_4^+	4.0	2.6	3.7	2.3	1.9	2.1

[a] April to December.
[b] January to April.

tural land, the strong influence of the wind direction on the wet deposition induces maximum values from a roughly southwestern direction (Fig. 8.2).

8.2 Deposition of SO_4^{2-}, NO_3^-, Cl^-, F^- and NH_4^+: Mean Values, Temporal Variations and Long-Term Trends

The analyses of rainwater during the period April 1987 to March 1991 revealed for the mean daily deposition the following values in mg/m^2d (Michaelis et al. 1992): sulphate 6.30; nitrate 7.07; chloride 2.68 and ammonium 3.00. Fluoride was only measured from April 1987 to August 1989. The mean value was 0.038 mg/m^2d.

Such long-term mean values, however, have in part only a limited usefulness insofar as temporal trends during the observation period may occur. This becomes obvious from Table 8.1 where the annual means for the total term of investigation are compiled. A distinct decreasing trend is perceptible in the case of sulphate. This development is due to the reduction of the sulphur dioxide emission in the West German States of the Federal Republic. From 1987 to 1988 this emission diminished from about 1.9 million tons per year to about 1.25 and further in 1989 to approximately 1.0 million tons per year (Bundesumweltministerium 1992). In the conurbation of Hamburg the decrease, on the basis of 1987 (32 430 t/a), may well have been 10 to 15 % (Umweltbehörde der Freien und Hansestadt Hamburg 1990). Weather factors have also contributed to the trend listed in Table 8.1 (Umweltbundesamt 1991). The year 1987 was characterized by a comparatively high amount of precipitation (cf. Fig. 7.7). Moreover, the winters in the period 1988 to 1990 were relatively mild. The mean winter temperatures at the investigation site amounted to 3.5 °C 1987/1988, 4.2 °C 1988/1989 and 4.0 °C 1989/1990.

In the case of nitrate, a significant decrease is not apparent. This finding is compatible with the trends of the NO_x emissions in the Federal Republic: in western Germany there was only a minor diminution, while in the east emissions stagnated at an approximately constant level (Bundesumwelt-

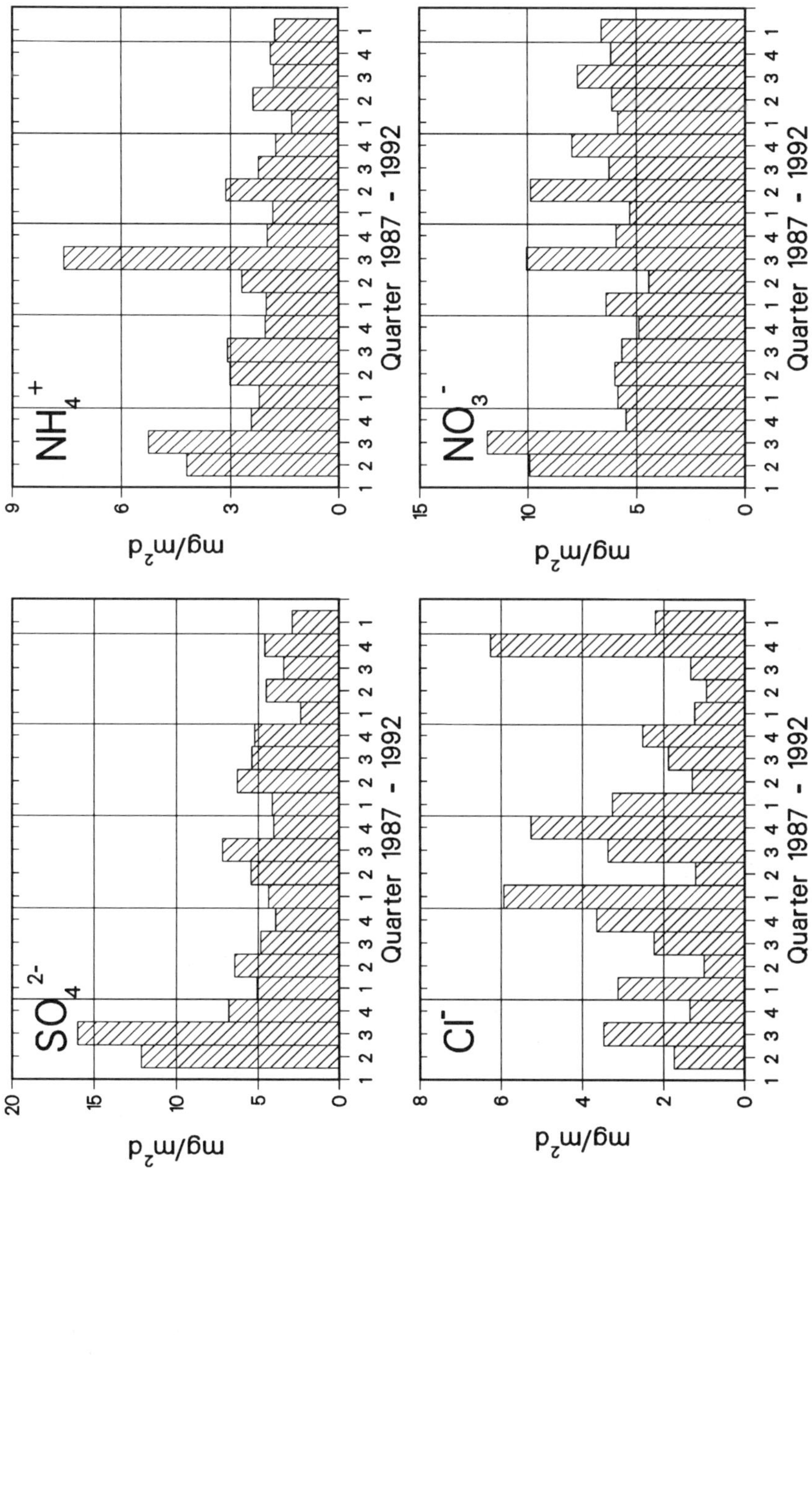

Fig. 8.3. Temporal variation in the daily wet deposition (mg/m²d) of the ions SO_4^{2-}, NO_3^-, Cl^- and NH_4^+ in a quarterly plot for the period 1987 to 1992

Table 8.2. Average daily deposition of ions in mg/m²d at Farchau/Ratzeburg (1987) and at South German measuring stations (1983/1986)

	SO_4^{2-}	NO_3^-	Cl^-	NH_4^+
Farchau/Ratzeburg[a] "Postturm"	11.5	9.0	2.3	4.0
Schönbuch I[b]	6.5–10.8	4.6–6.8	1.8–5.3	1.8–2.1
Schönbuch II[c]	9.1–15.7	6.5–9.9	2.0–5.2	2.6–3.4
Rotenfels[b]	9.5–15.3	8.1–13.3	3.5–6.5	1.9–2.9
Freudenstadt[b)]	7.8–18.1	6.5–10.3	3.1–6.9	1.8–2.2

[a] Michaelis et al. (1992).
[b] Adam et al. (1987).
[c] Baumbach et al. (1987).

ministerium 1992). For all other ions there is also no significant trend discernible, if the weather factors and the experimental limits of error are taken into consideration.

The deposition of ions shows distinct seasonal variations (Fig. 8.3). In the case of SO_4^{2-}, NO_3^- and NH_4^+ similarities to the course of the amount of precipitation (Fig. 7.7) and that of important trace elements (Fig. 7.8) are obvious. Chloride, however, exhibits a different behaviour with maxima occurring mostly during the windy winter periods which supports the assumption that the sea is a perceptible source of Cl^- ions.

When comparing the data with findings from southwest German investigation sites, it is recommendable to select measuring periods which overlap or are at least adjacent, in order to minimize the influence of temporal trends. Therefore, in Table 8.2, the results of the present study from 1987 are compared with literature data from measurements performed in the directly preceding years (Adam et al. 1987; Baumbach et al. 1987). A short survey of the southwest German sites considered has already been given in Section 7.2.2. As can be seen from the Table 8.2, the deposition on the whole does not show serious differences. A striking result is the higher input of Cl^- measured at the southern stations. This could be an indication of stronger anthropogenic emissions in these regions. Another remarkable outcome is the high deposition of NH_4^+ at the "Postturm" site. Most probably, the main causes are the large, intensively utilized agricultural areas and the mass cattle keeping in the neighbourhood.

8.3 pH Values and H^+-Ion Deposition

The results of the pH measurements are summarized in Table 8.3. As can be seen, the seasonal variations are not pronounced and no significant long-term trend is perceptible. For comparison, data from the official monitoring

Table 8.3. pH values of rainwater samples. Quarterly and annual means

Quarter	1987	1988	1989	1990	1991	1992	Average
I	-	4.15	4.34	4.45	4.32	4.00	4.25
II	4.37	4.06	4.14	4.13	4.14	-	4.17
III	4.19	4.34	4.27	4.27	4.08	-	4.23
IV	4.01	4.29	4.22	4.21	4.28	-	4.20
Annual mean "Postturm"	4.19[a]	4.21	4.24	4.27	4.21	-	-
Annual mean Hahnheide	4.25	4.49	4.64	4.75	4.63	-	-

[a] April to December only.

Table 8.4. Quarterly and annual mean values of the H^+-deposition in mg/m^2d

Quarter	1987	1988	1989	1990	1991	1992	Average
I	-	0.22	0.06	0.07	0.05	0.16	0.11
II	0.11	0.10	0.09	0.14	0.16	-	0.12
III	0.26	0.08	0.14	0.12	0.11	-	0.14
IV	0.16	0.11	0.10	0.13	0.10	-	0.12
Annual mean "Postturm"	0.18[a]	0.13	0.10	0.12	0.11	-	-
Annual mean Hahnheide	0.15	0.093	0.063	0.074	0.063	-	-

[a] April to December only.

programme in Schleswig-Holstein (Landesamt für Wasserhaushalt und Küsten Schleswig-Holstein 1995) have also been included. They refer to a measuring site about 20 km to the west of the "Postturm" location (Hahnheide) and are based on fortnightly sampling. These data are on the whole greater than those measured in the present study. The main reason is most probably the even higher content of ammonium at the Hahnheide site and the associated stronger neutralization capacity. The data also suggest a slight increase in the pH values. Such an encouraging trend, however, has to be assessed with care since the respective basic constituents must be taken into account as well. This might also be the cause of the fact that the results obtained at the "Postturm" site do not exhibit any discernible tendency beyond the experimental limits of error.

The quarterly and annual means of the H^+-ion deposition are compiled in Table 8.4. Fluctuations are due solely to the influence of meteorological conditions and, as expected from Table 8.3, no long-term trend becomes apparent. Again, data from the above-mentioned monitoring programme (Landesamt für Wasserhaushalt und Küsten Schleswig-Holstein 1995) have been included.

References

Adam K, Evers FH, Littek T (1987) Ergebnisse niederschlagsanalytischer Untersuchungen in süddeutschen Waldökosystemen 1981–1986. In: Projekt Europäisches Forschungszentrum für Maßnahmen zur Luftreinhaltung (PEF) im Kernforschungszentrum Karlsruhe (ed), KfK-PEF 24

Baumbach G, Dröscher F, Mikisch E (1987) Staub- und Niederschlagsuntersuchungen in Wäldern. Bericht Nr. 9, Universität Stuttgart

Bundesumweltministerium (ed)(1992) Umweltschutz in Deutschland - Nationalbericht der Bundesrepublik Deutschland für die Konferenz der Vereinten Nationen über Umwelt und Entwicklung in Brasilien. Economica Verlag, Bonn

Landesamt für Wasserhaushalt und Küsten Schleswig-Holstein (ed)(1995) Ein Jahrzehnt Beobachtung der Niederschlagsbeschaffenheit in Schleswig-Holstein 1985–1994. Landesamt für Wasserhaushalt und Küsten Schleswig-Holstein, Kiel

Michaelis W, Pepelnik R, Theopold F, Rademacher P (1992) Deposition atmosphärischer Spurenstoffe und Stoffflüsse im Ökosystem Wald. In: Michaelis W, Bauch J (eds) Luftverunreinigungen und Waldschäden am Standort "Postturm", Forstamt Farchau/Ratzeburg. GKSS Forschungszentrum Geesthacht, GKSS 92/E/100, pp 11–59

Umweltbehörde der Freien und Hansestadt Hamburg (ed)(1990) Luftreinhaltung in Hamburg - Sachstandsbericht 1990

Umweltbundesamt (ed)(1991) Jahresbericht 1991, Berlin, pp 210–211

9 Trace Elements in Size-Fractionated Particulates: Concentrations and Dry Deposition into a Forest Ecosystem

9.1 Differential Trace Element Concentrations

Considering the dependence of the deposition velocity on the particle-size (cf. Fig. 6.2), exact knowledge of the concentration distribution over the size-fractions is an essential precondition for a correct determination of dry deposition. The relative contributions of the individual size fractions to the total flux vary from element to element, since the concentration distributions can be quite different. Here, the origin plays an important part. Elements of predominantly anthropogenic origin, for instance Cd or Pb, show a rise in the concentration in air with decreasing aerodynamic diameter. This is the outcome of the long-range transport. On the other hand, elements of essentially geogenic origin, such as Ca and Fe, exhibit a contrary behaviour: the concentration decreases with diminishing diameter; nearby sources markedly contribute to the element content in the particulate phase. These facts are illustrated in Fig. 9.1, where the differential trace element concen-

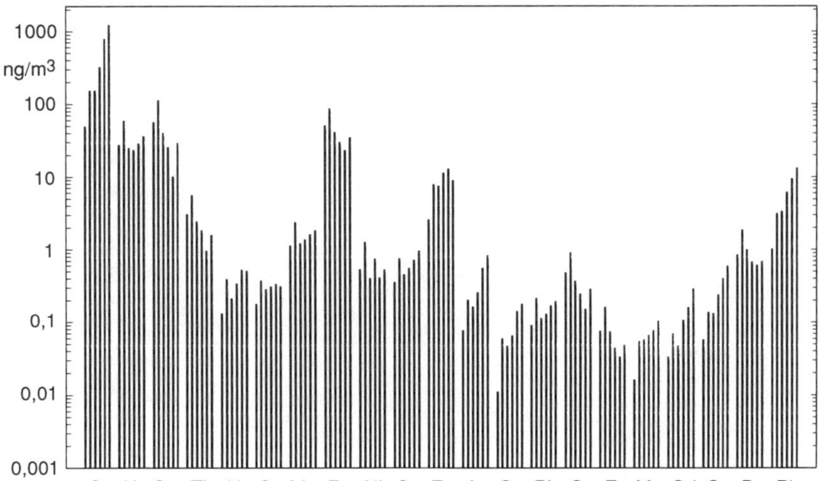

Fig. 9.1. Airborne particulates: distributions of the element concentrations in air (ng/m^3) as a function of the particle size. Decreasing aerodynamic diameter from *left to right*. Fractions according to Section 5.1.3

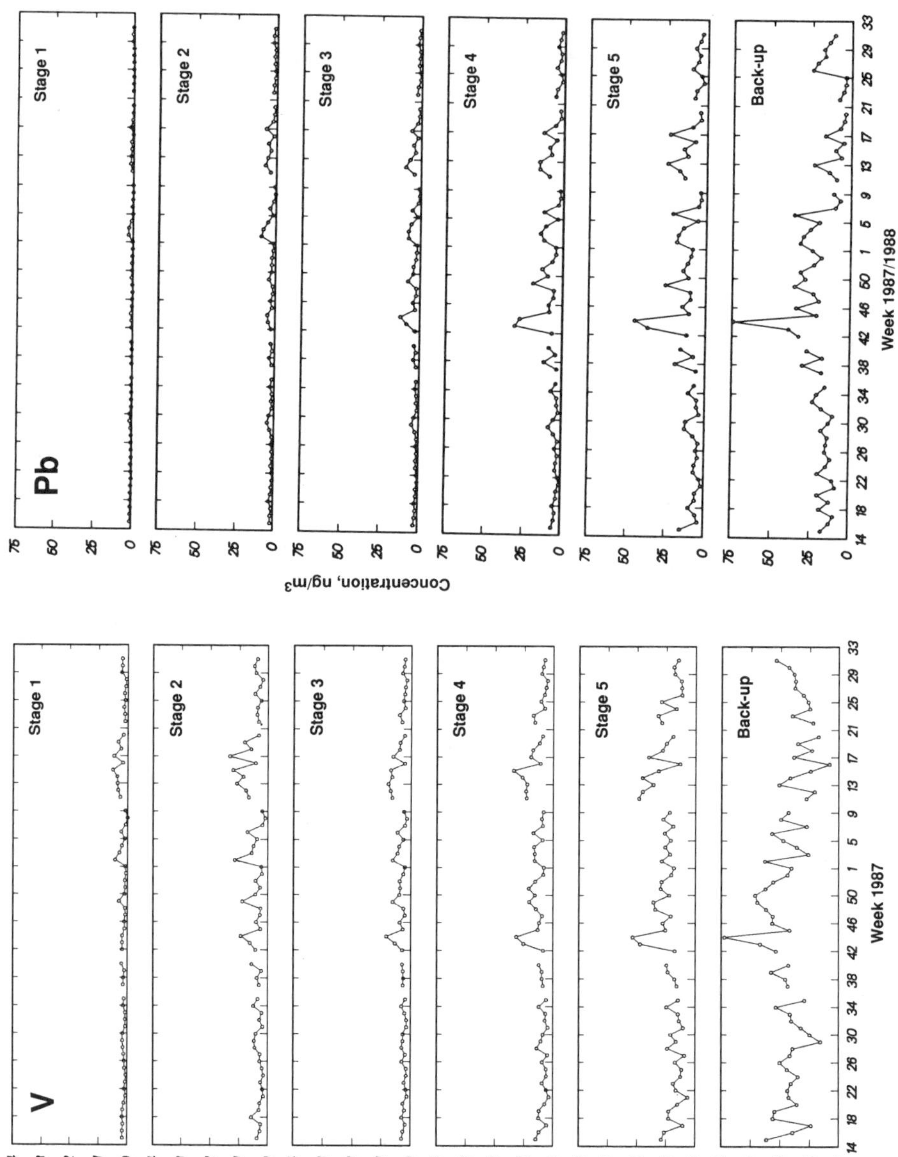

Fig. 9.2. Temporal variations in the differential concentrations of vanadium and lead in air bonded to airborne particulates. Measurement using a 5-stage impactor with backup filter during the period 15th calendar week 1987 to 31st week 1988

trations are plotted with decreasing particle diameter from left to right according to the six fractions of the impactor used (cf. Sect. 5.1.3). The diagram also demonstrates the excellent detection sensitivity of the applied procedures which is in the order of pg/m^3.

The temporal variations of the differential concentrations are very pronounced and a proportionality cannot be assumed in general. This is shown in Fig. 9.2 for the elements V and Pb. Vanadium is a trace element which originates mainly from the combustion of oil. This may be the reason for elevated values during the winter periods.

9.2 Dry Deposition of Trace Elements

9.2.1 Meteorological Aspects

As is the case for the trace element concentrations in rainwater, the total concentrations of particle-bonded elements in air also show a clear dependence on the wind direction. This is exemplified in Fig. 9.3 for sulphur and lead on the basis of data from 1988. They are plotted in mg/m^3 and µg/m^3, respectively. The total concentration has to be understood as the sum over all size fractions, the number of which is six in the present study. Again, it should be stressed that the plots in Fig. 9.3 and also those in Fig. 9.5 have to be comprehended in a more qualitative than quantitative manner, since only samples which could be related to rather stable meteorological conditions were taken into account.

In contrast to the above-mentioned concentrations, a deviating behaviour occurs between dry and wet deposition. The reason is that the frequency distribution of the wind direction exhibits less anisotropy than the distribution of the amount of precipitation. This becomes evident when comparing Fig. 9.4 with Fig. 7.5 in Section 7.2.1. The influence of the friction velocity does not basically alter the conditions. Important consequences result from these considerations. The geographical distributions of the sources are not identical for wet and dry deposition. In the latter process, the pollutant transport is not so much restricted to a rather limited sector as is the case in the former one. In particular, sources southeast of the measuring station contribute more strongly, with the consequence that the dry deposition of some important elements, at least during the first half of the investigation period, shows a lower decrease than the wet deposition and this is due to the inferior environmental policy in the former German Democratic Republic. By virtue of the magnitude of dry deposition, this also affects the trend of the total flux of the respective elements. Concrete data on these facts will be presented in the subsequent sections and in Chapter 10.

In Fig. 9.5, findings from 1988 on the dry deposition of the elements sulphur and lead as a function of the wind direction are displayed. A comparison with Fig. 7.6 substantiates the above statements.

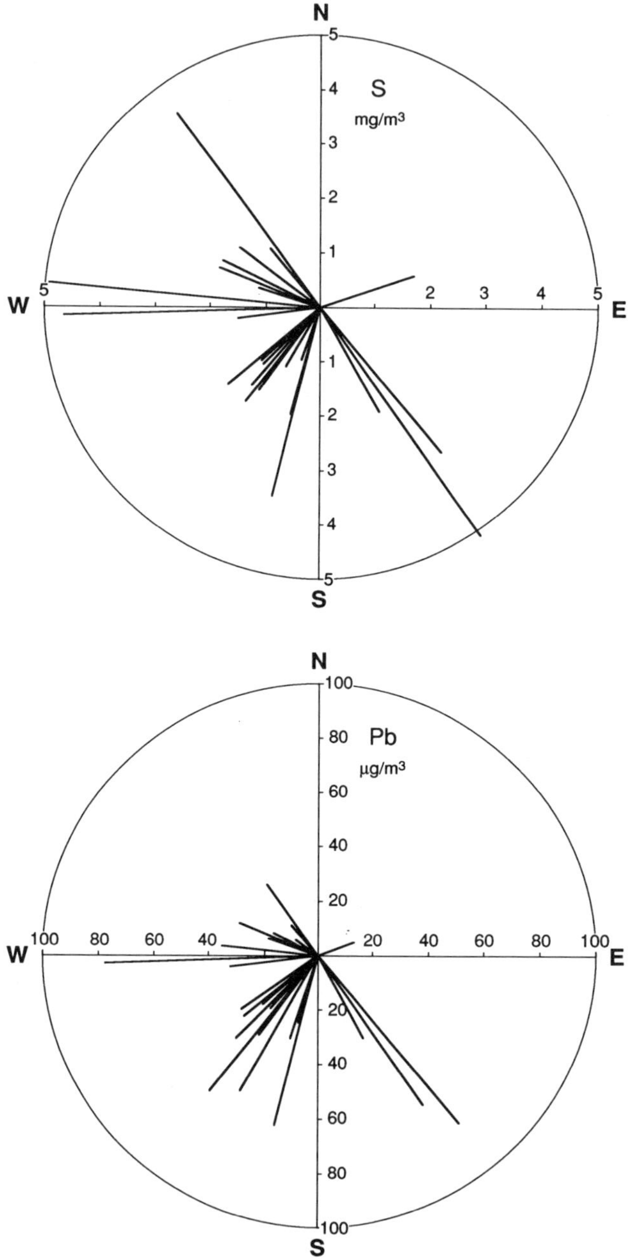

Fig. 9.3. Total concentration of particle-bonded sulphur (mg/m^3) and lead (µg/m^3) in air vs. wind direction. Data from 1988

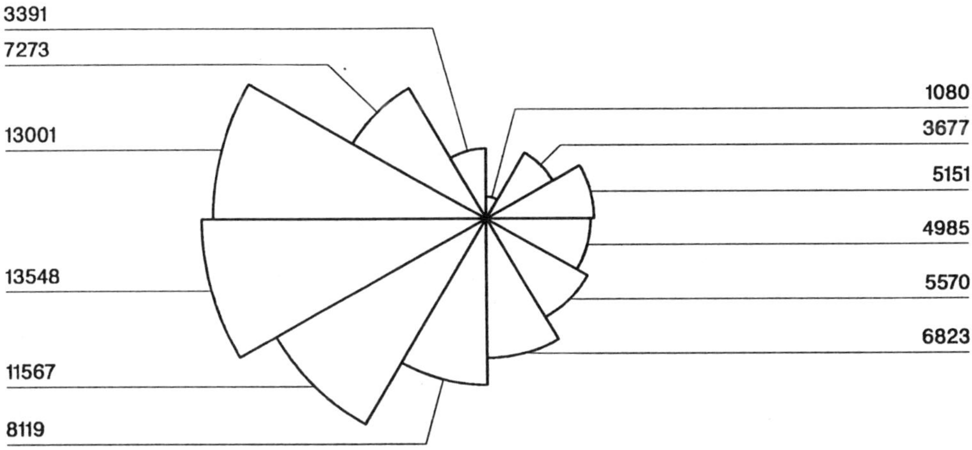

Fig. 9.4. Frequency distribution of the wind direction (30° sectors). Long-term means over the period 1987 to 1992. The *numbers* indicate the quantity of measured values

Table 9.1. Weighted long-term average values of the daily dry deposition of trace elements. Measuring period April 1987 to March 1991

Element	µg/m²d	Element	µg/m²d
S	7460 ± 2000	Ni	10.5 ± 2.5
K	765 ± 215	Cu	20 ± 6
Ca	1850 ± 560	Zn	126 ± 34
Ti	66 ± 20	As	9.5 ± 2.5
V	14 ± 3	Se	2.7 ± 0.7
Cr	6.9 ± 2.1	Mo	1.4 ± 0.4
Mn	35 ± 10	Cd	2.3 ± 0.6
Fe	935 ± 285	Pb	119 ± 31

9.2.2 Long-Term Average Values

The investigation of the dry deposition was one of the focal points of the present study, since this process makes up the greater component of the total element flux into a forest ecosystem (Michaelis et al.1985a,b, 1988, 1989, 1992a,b,c; Michaelis 1986a,b, 1988, 1991; Michaelis and Prange 1988; Hertlein 1990; Rademacher et al.1992; Pepelnik et al.1993). After several years of intensive effort to improve the methodical fundamentals, a sufficient accuracy of the experimental results has in the meantime been achieved. The consistency of the data, for instance regarding the balance of the element fluxes (cf. Chap. 11), supports this conclusion. Nevertheless, the error limits are still markedly higher than for the determination of the wet deposition. This is above all due to the remaining uncertainties in the

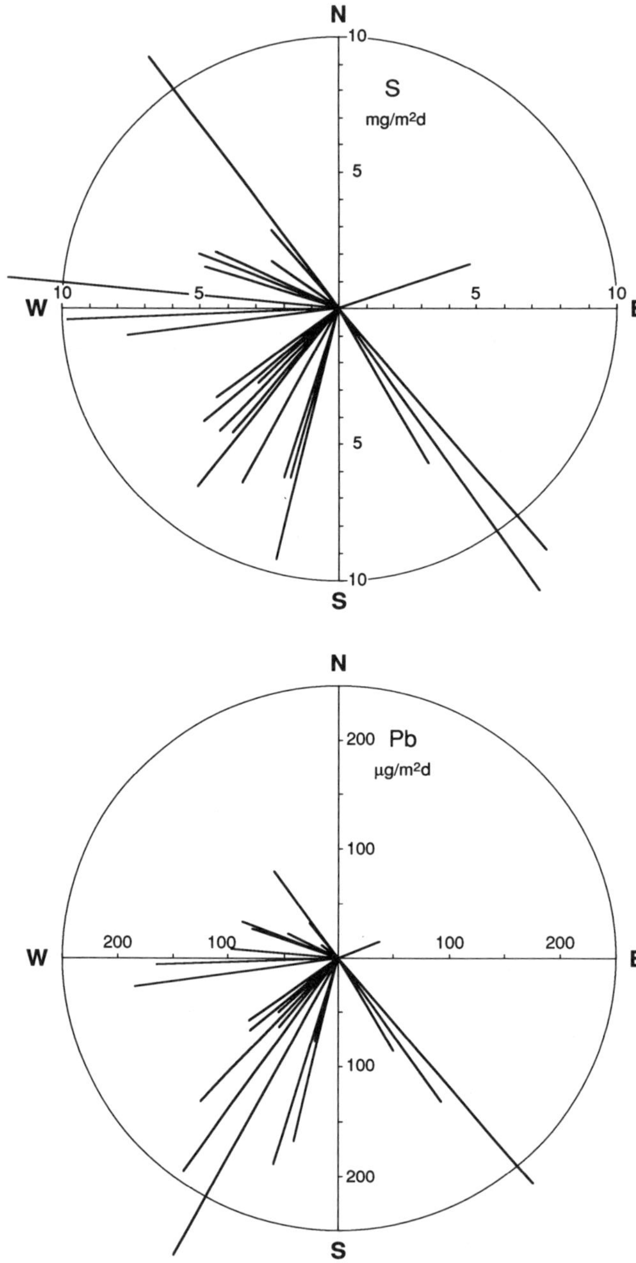

Fig. 9.5. Dry deposition of sulphur (mg/m²d) and lead (µg/m²d) vs. wind direction. Data from 1988

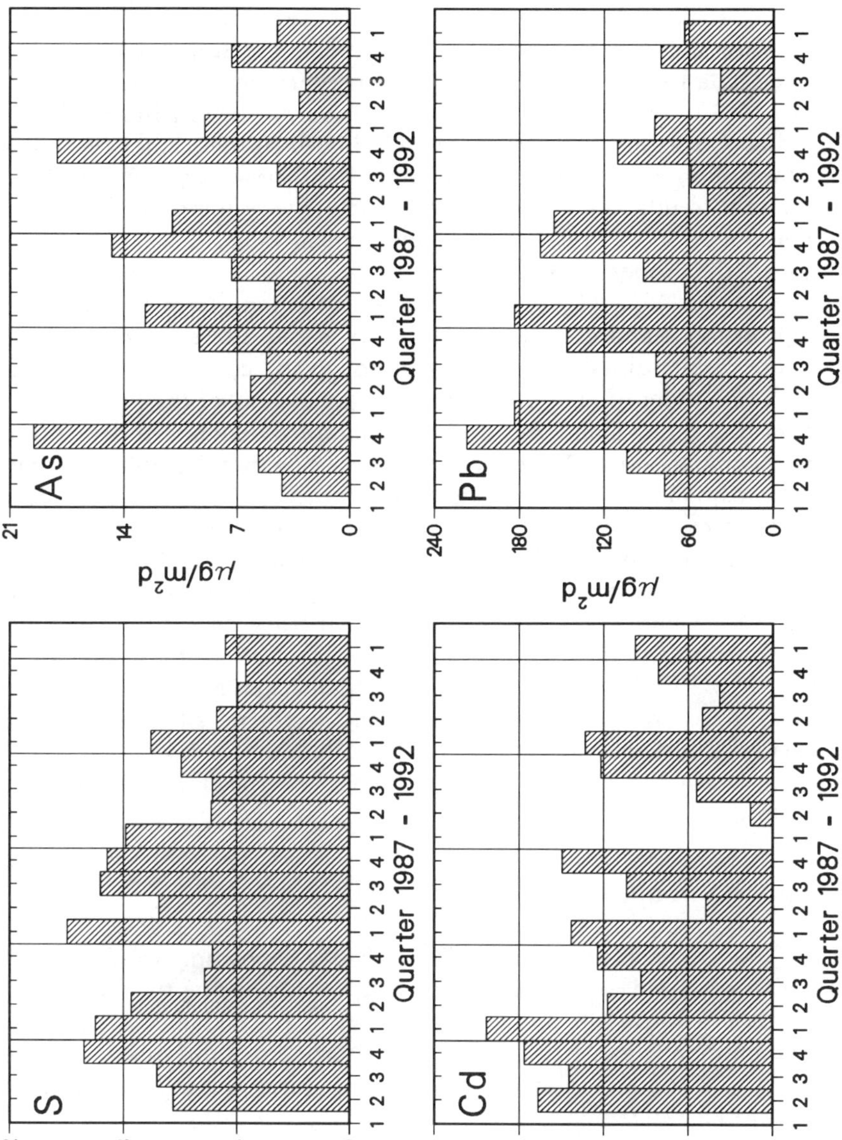

Fig. 9.6. Seasonal variations in the daily dry deposition of S, As, Cd and Pb. Quarterly plot from 1987 to 1992

deposition velocity. Other possible sources of errors are the degreee of density and details of the composition of the spruce stand.

Results for the dry deposition of 16 elements are summarized in Table 9.1. The data listed represent weighted long-term averages over the period April 1987 to March 1991. These values have, on the one hand, the advantage that short-time fluctuations are eliminated so that rather representative information is furnished. On the other hand, however, they hide possible temporal trends, the knowledge of which is very important for environmental policy. In this respect, the annual mean values discussed in Section 9.2.4 are the proper basis.

If the data of Table 9.1 are compared with the long-term mean values measured on the island of Pellworm (Michaelis and Stößel 1986), it turns out that the results differ considerably, although the concentrations in air are of the same order of magnitude. This is the outcome of the pronounced filter effect of a forest stand and the correspondingly enhanced deposition velocity. A comparison of Tables 7.2 and 9.1 reveals that the total flux into the forest ecosystem is to a large extent determined by dry deposition. It is true that the wet deposition introduces toxic elements in a chemical form most favourable for the uptake by plants, but the acidity of the rain gives rise to the mobilization and availability also of constituents of deposited particulates (cf. Chap. 11). Thus, the extent of the dry deposition can substantially contribute to the high heavy metal concentrations in the soil solutions and in the fine roots of the trees (Rademacher et al.1988; see also Chap. 11). Indeed, the pH values (KCl) at the "Postturm" site amounted to about 2.8 in the horizon 0 to 5 cm, between 3.2 to 3.5 in the range 5 to 25 cm and up to 4.2 in the deeper mineral soil (25 to 100 cm) (Rademacher et al.1992).

9.2.3 Seasonal Variations

In Fig. 9.6, the seasonal variations of the daily dry deposition are shown for the elements S, As, Cd and Pb. Maxima mostly occur during the winter period. This is in contrast to the wet deposition of trace elements and ions which predominantly exhibits maxima in the second or third quarter of the year (cf. Figs. 7.8 and 8.3) in accordance with the variation in the amount of precipitation (Fig. 7.7).

9.2.4 Annual Mean Values and Long-Term Trends

A summary of the results on the annual mean values is given in Table 9.2. As can be seen, the elements sulphur, zinc, arsenic, cadmium and lead exhibit a distinct decreasing tendency. Such a trend is not discernible in the case of antimony. On the basis of Fig. 7.4, one might have expected a decrease for this element, too.

Table 9.2. Annual average values of the daily dry deposition of trace elements. All data in μg/m²d

Element	1987[a])	1988	1989	1990	1991	1992[b])
S	7570	6520	8590	5490	5690	4800
K	670	816	919	575	656	500
Ca	1160	1650	1660	1920	2250	950
Ti	50	80	76	49	55	37
V	13.6	14.8	14.6	11.4	14.2	14.6
Cr	5.7	7.4	7.8	7.1	6.3	5.7
Mn	29	39	39	25	25	20
Fe	834	1075	1040	671	690	480
Ni	8.0	12.6	9.5	9.0	7.1	6.2
Cu	18.3	21.2	22.1	14.9	18.7	14.3
Zn	141	141	125	(77)	82	82
As	9.8	8.6	8.9	8.3	5.9	5.0
Se	2.1	2.3	3.0	2.6	2.8	3.8
Rb	3.0	3.6	3.6	2.3	2.3	1.8
Sr	9.8	13.3	11.9	8.1	7.2	6.3
Y	0.2	0.7	0.5	0.3	(0.8)	0.2
Zr	2.7	4.6	4.7	4.3	2.9	2.1
Nb	0.1	0.3	0.3	0.1	(0.4)	0.2
Mo	1.3	1.5	1.6	0.7	1.0	1.2
Cd	2.7	2.2	1.7	1.3	1.3	1.5
Sn	5.5	6.5	7.7	4.3	3.4	4.3
Sb	2.5	3.3	3.3	2.5	3.0	3.3
Ba	18.5	26.1	21.4	13.7	11.3	9.0
Pb	135	121	118	77	60	69

[a] April to December only
[b] January to April only

It is conspicuous that the dry deposition, in particular of S and Pb, shows a less pronounced trend than the wet deposition and that the decrease clearly started later (cf. Table 7.5). These findings can be explained by the different influence that the meteorology has on the two deposition modes, as has already been discussed in Section 9.2.1. In 1987, sulphur dioxide emission in the former German Democratic Republic was the highest of all European states, amounting to about 5.5 million tons/year (cf. Chap. 12). The comparative value for the old Federal Republic of Germany was about 1.9 million tons/year at that time (Bundesumweltministerium 1992), and that with an area and a population greater by approximately a factor of 2.3 and 3.5, respectively. In the period 1987 to 1989, the emission of sulphur dioxide in the German Democratic Republic decreased only slightly, however,it has markedly declined since 1990, at the beginning above all due to numerous industrial shutdowns (see also Chap. 12).

In the case of lead, the circumstances were similar. While in the West unleaded petrol had already been introduced before the beginning of the investigation period, this happened in the East only after the German unification. Moreover, the motor traffic in the East increased considerably after this date. The trend of the dry deposition of zinc, arsenic and cadmium is probably first of all determined by sources in the western vicinity (cf. Sect. 7.1).

References

Bundesumweltministerium (ed)(1992) Umweltschutz in Deutschland - Nationalbericht der Bundesrepublik Deutschland für die Konferenz der Vereinten Nationen über Umwelt und Entwicklung in Brasilien. Economica Verlag, Bonn

Hertlein F (1990) Untersuchungen zur Anwendung der Gradientenmethode auf luftgetragene Partikel. Diplomarbeit, Fachbereich Angewandte Naturwissenschaften, Fachhochschule Lübeck

Michaelis W (1986a) Naß- und Trockendeposition von Schwermetallen. In: Klose W, Leßmann E (eds) Bodenschutz - Lösung durch Technik, ENVITEC 86. Vulkan-Verlag, Essen, pp 36–41

Michaelis W (1986b) Multielement analysis of environmental samples by total-reflection X-ray fluorescence spectrometry, neutron activation analysis and inductively coupled plasma optical emission spectroscopy. Fresenius Z Anal Chem 324:662–671

Michaelis W (1988) Experimental studies on dry deposition of heavy metals and gases. In: van Dop H (ed) Air pollution modeling and its application VI, vol 11. Plenum, New York, pp 61–74

Michaelis W (1991) Trace analysis of geological and environmental samples. In: Störr M, Henning KH, Adolphi P (eds) Proc 7th EUROCLAY Conf, Dresden '91, vol II. Ernst-Moritz-Arndt Universität Greifswald, pp 761–766

Michaelis W, Prange A (1988) Trace analysis of geological and environmental samples by total-reflection X-ray fluorescence spectrometry. Nucl Geophys 2(4):231–245

Michaelis W, Stößel RP (1986) Untersuchungen zur Schwermetalldeposition auf der Insel Pellworm - ein Beitrag zur Ermittlung des atmosphärischen Schadstoffeintrags in die Nordsee. In: GKSS Forschungszentrum Geesthacht, GKSS Jahresbericht 1986, pp 8–23

Michaelis W, Fanger HU, Niedergesäß R, Schwenke H (1985a) Intercomparison of the multielement analytical methods TXRF, NAA and ICP with regard to trace element determinations in environmental samples. In: Sansoni B (ed) Instrumentelle Multielementanalyse. VCH Verlagsgesellschaft, Weinheim, pp 693–709

Michaelis W, Prange A, Knoth J (1985b) Applications of total-reflection X-ray fluorescence in multi-element analysis. In: Sansoni B (ed) Instrumentelle Multielementanalyse. VCH Verlagsgesellschaft, Weinheim, pp 269–289

Michaelis W, Schönburg M, Stößel RP (1988) Trocken- und Naßdeposition von Schwermetallen und Gasen. In: Bauch J, Michaelis W (eds) Das Forschungsprogramm Waldschäden am Standort "Postturm", Forstamt Farchau/Ratzeburg. GKSS Forschungszentrum Geesthacht, GKSS 88/E/55, pp 19–59

Michaelis W, Schönburg M, Stößel RP (1989) Deposition of atmospheric pollutants into a North German forest ecosystem. In: Georgii HW (ed) Mechanisms and effects of pollutant-transfer into forests. Kluwer, Dordrecht, pp 3–12

Michaelis W, Pepelnik R, Theopold F, Rademacher P (1992a) Deposition atmosphärischer Spurenstoffe und Stoffflüsse im Ökosystem Wald. In: Michaelis W, Bauch J (eds) Luftverunreinigungen und Waldschäden am Standort "Postturm", Forstamt Farchau/Ratzeburg. GKSS Forschungszentrum Geesthacht, GKSS 92/E/100, pp 11–59

Michaelis W, Pepelnik R, Prange A (1992b) Application of TXRF in environmental research. In: Barret CS, Gilfrich JV, Huang TC, Jenkins R, McCarthy GJ, Predecki PK, Ryon R, Smith DK (eds) Advances in X-ray analysis, vol 35B. Plenum Press, New York, pp 953–958

Michaelis W, Pepelnik R, Rademacher P, Riebesell M (1992c) Transfer of atmospheric pollutants into a forest ecosystem. In: Teller A, Mathy P, Jeffers JNR (eds) Responses of forest ecosystems to environmental changes. Elsevier, London, pp 596–597

Pepelnik R, Erbslöh B, Michaelis W, Prange A (1993) Determination of trace element deposition into a forest ecosystem using total-reflection X-ray fluorescence. Spectrochim Acta 48B(2):223–229

Rademacher P, Bauch J, Michaelis W (1988) Einfluß der Elementkonzentration der Bodenlösung auf den Elementgehalt in gesunden und geschädigten Fichten des Standortes "Postturm". In: Bauch J, Michaelis W (eds) Das Forschungsprogramm Waldschäden am Standort "Postturm", Forstamt Farchau/Ratzeburg. GKSS Forschungszentrum Geesthacht, GKSS 88/E/55, pp 215–254

Rademacher P, Ulrich B, Michaelis W (1992) Bilanzierung der Elementvorräte und Elementflüsse innerhalb der Ökosystemkompartimente Krone, Stamm, Wurzel und Boden eines belasteten Fichtenbestandes am Standort "Postturm". In: Michaelis W, Bauch J (eds) Luftverunreinigungen und Waldschäden am Standort "Postturm", Forstamt Farchau/Ratzeburg. GKSS Forschungszentrum Geesthacht, GKSS 92/E/100, pp 149–186

10 Total Deposition of Trace Elements

In contrast to grass surfaces where the wet deposition, depending on wind conditions and hence friction velocity, contributes from one-half to two-thirds of the total deposition (Michaelis et al.1989, 1990), the total flux into forests is predominantly determined by the dry deposition. As a consequence, in this case, rainfall studies alone do not yield representative insight into the atmospheric impact. Figure 10.1 illustrates the relative portions of the two deposition modes for the elements sulphur, arsenic, cadmium and lead plotted quarterly for several years of the investigation period. The partition differs from element to element and in part also from year to year. In each case, a distinct seasonal variation occurs due to the meteorological influence (cf. Chaps. 7 and 9).

Since the dry deposition of some elements at the "Postturm" site is to no small degree also controlled by emission sources in eastern Germany (Sects. 9.2.1 and 9.2.4), the total impact decreased with a delay and not to the extent which might have been expected on the basis of the emission reductions in the west. Air quality measures in the new German States are therefore equally relevant for westerly regions. The temporal development of the total deposition relative to the value of 1987 is represented in Fig. 10.2 for ten ecologically significant elements. It can be seen in this diagram that there is a decreasing trend in the case of sulphur, nickel, copper, zinc, arsenic, cadmium and lead, while in consideration of the experimental limits of error for the elements vanadium, chromium and antimony, no tendency of any kind is discernible.

According to the Tables 7.5 and 9.2, the total deposition of lead in 1987/1988 amounted to 141 $\mu g/m^2$ d. This value compares quite well with data reported for the Teutoburg Forest (Godt 1985) where the total flux was 113 $\mu g/m^2$ d inside the stand and 190 $\mu g/m^2$ d at its edge.

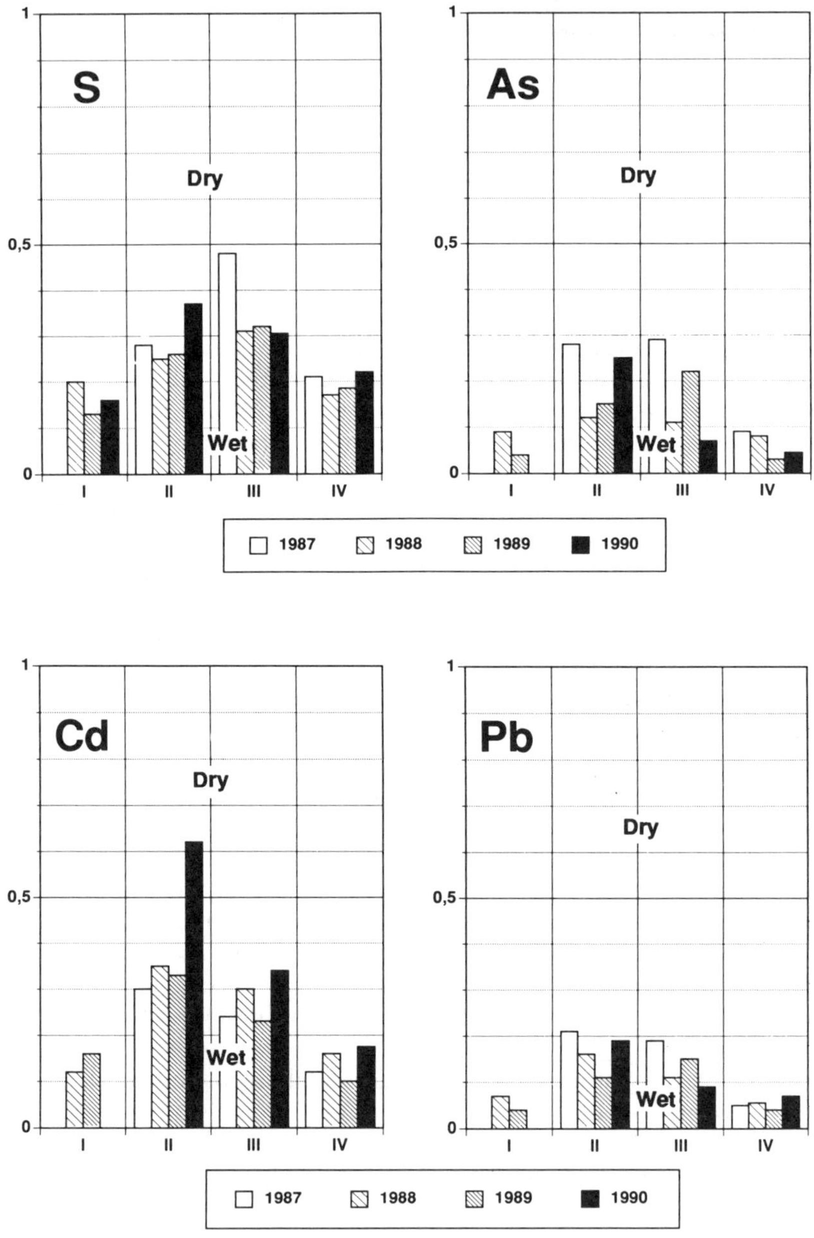

Fig. 10.1. Relative contributions of wet and dry deposition of S, As, Cd and Pb to the total flux in a quarterly plot (*I* to *IV*) during the period 1987 to 1990

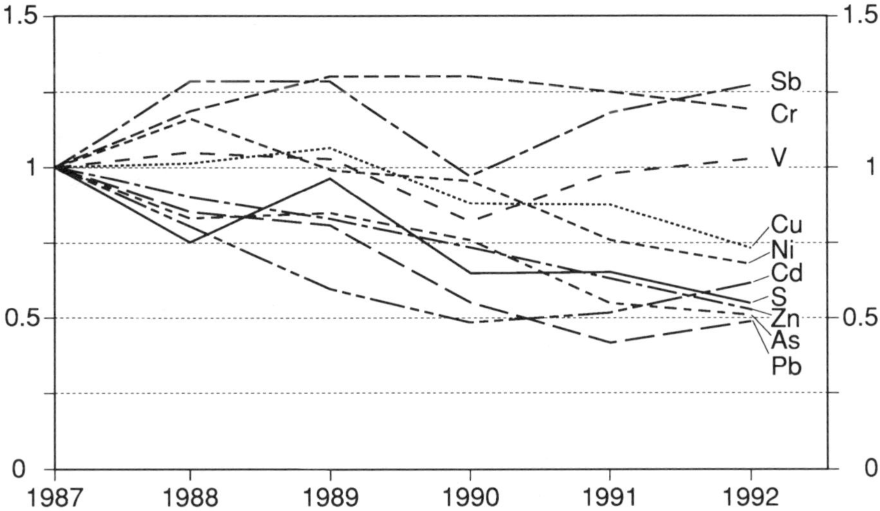

Fig. 10.2. Temporal variation in the annual mean values of the total deposition in relation to the basis 1987 for ten ecologically relevant trace elements

References

Godt J (1985) Schwermetallbelastung des Teutoburger Waldes südwestlich der Stadt Detmold. Bielefelder Ökol Beitrag 1:7–16

Michaelis W, Schönburg M, Stößel RP (1989) Schadstofftransfer in der Grenzschicht Atmosphäre-Vegetation. In: Arbeitsgemeinschaft der Großforschungseinrichtungen (ed) Wechselwirkung Atmosphäre-Biosphäre. Thenée Druck, Bonn, pp 29–33

Michaelis W, Pepelnik R, Rademacher P, Riebesell M (1990) Wechselwirkung zwischen Luftschadstoffen und Vegetation. GKSS Forschungszentrum Geesthacht, GKSS-Jahresbericht 1990, pp 42–55

11 Element Supply and Element Fluxes in the Forest Ecosystem Compartments

11.1 Element Supply in the Mineral Soil and the Humus Layer

For the official assessment of forest decline the trees are primarily evaluated with regard to visible symptoms concerning their vitality. However, in the case of spruce, needle chlorosis and needle loss, which are used as essential criteria, are often not in accordance with the results of special investigations. In particular, growth (Athari and Kramer 1983; Bauch et al. 1986) and nutrition-physiological parameters (Fedderau-Himme et al. 1984; Horn and Zech 1987) frequently do not correlate with the state of the needles. Therefore, numerous studies were initiated with the goal of taking into account as many physiological and structural properties of the diverse tree tissues as possible for a better characterization of trees belonging to the different damage classes (Bauch et al. 1979; Berchthold et al. 1981; Rehfuess 1983; Bosch et al. 1983; Schütt et al. 1983; Fedderau-Himme et al. 1984; Jüttner 1985; Frenzel and Christmann 1986; Rademacher 1986; Schulz and Behnke 1986; Oren et al. 1989; Riederer 1989; Suske and Acker 1989). For identifying the processes leading to forest decline particular attention was focused on the effects that atmospheric pollutants have on the plant metabolism, on the carbohydrate supply via the needles and on the element supply through the fine roots (see, among others, Marschner and Richter 1974; Bauch and Schröder 1982; Landolt 1982; Ulrich 1983; Elstner and Osswald 1984; Murach 1984; Krivan et al. 1986; Stienen 1985; Oberwinkler et al. 1986; Wild and Bode 1986; Hüttl 1987; Isermann 1987; Eschrich 1988; Klemm 1989; Lange et al. 1989; Oren and Zimmermann 1989; Schneider et al. 1989). Extensive studies on the chemical soil conditions indicate that at many sites the soil solutions constitute an insufficient basis for the nutrition of the trees (Ulrich 1981; Gehrmann et al. 1984; Schulte-Bisping and Murach 1984; Hildebrand 1985; Hartmann et al. 1989; Horn et al. 1989; Kaupenjohann 1989; Kaupenjohann et al. 1989; Schulze et al. 1989). Focal points of the trace analytical investigations at the "Postturm" site were the disclosure of the relations between the element concentrations in the soil solutions of diverse soil horizons and those in the fine roots and other tree compartments as well as the determination of element fluxes which are important for the element supply to the ecosystem (Rademacher et al. 1988; Panten 1990;

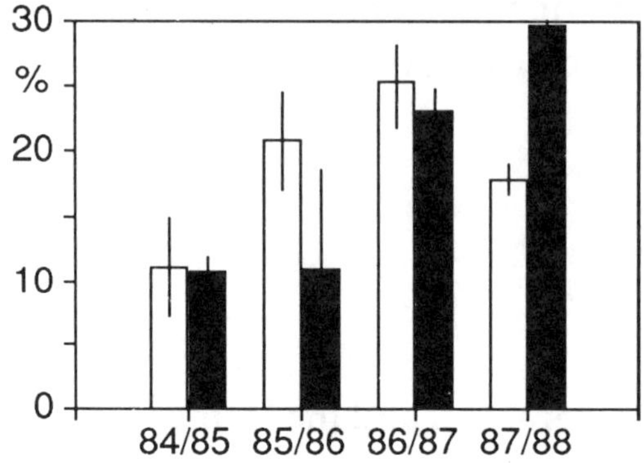

Fig. 11.1. Percentage of the soil moisture in the horizon O_{fh}/A (+8 to –5 cm) during the winter periods from 1984 to 1988. *Open bars* Damage class 0; *full bars* damage class 3

Schultz 1991; Michaelis et al. 1991, 1992a,b,c; Rademacher et al. 1992). In these studies, seasonal variations of the element contents as well as collectives of both healthy and declining trees were also considered. All data given in this chapter on the element content of organic matrices refer to the dry weight of the biomass.

The concentration of available elements in the soil solution depends not only on the chemical soil quality, but also shows a seasonal variation as a function of the soil moisture which exhibits maxima during the winter periods due to decreasing transpiration and evaporation. A comparison of the soil moisture during various periods of vegetation rest also reveals marked differences from year to year. This is exemplified in Fig. 11.1 where data obtained for the O_{fh}/A horizon (+8 to –5 cm) during four successive winters are plotted. In each case, the samples were taken in February and the moisture content was determined by centrifugation. Allowance was made for fine root spheres of both healthy and damaged trees (damage classes 0 and 3, respectively). For the classification, the definitions compiled by Uhlmann et al. (1989) were taken as a basis. In connection with Fig. 11.1, it must also be considered that by analogy with the soil moisture the pH value increased from 2.7 in February 1985 to 3.6 in 1986; in 1987 and 1988 it amounted to 3.4. These results substantiate a severe acidification of the soil. In the winter of 1984/1985, the state of the crown compartment of the trees showed a distinct deterioration.

The concentrations of nine important elements in the soil solution of the O_{fh}/A (+8 to –5 cm) horizon during the successive winter periods mentioned above are plotted in Fig. 11.2. As can be seen, a decreasing trend emerges for the macronutrients K, Ca, Mg, P and S. The concentrations of Fe, Mn and in particular Pb were found to be rather high. Al is only of limited importance in the horizon considered here. This outcome changes drastically, if deeper layers are examined.

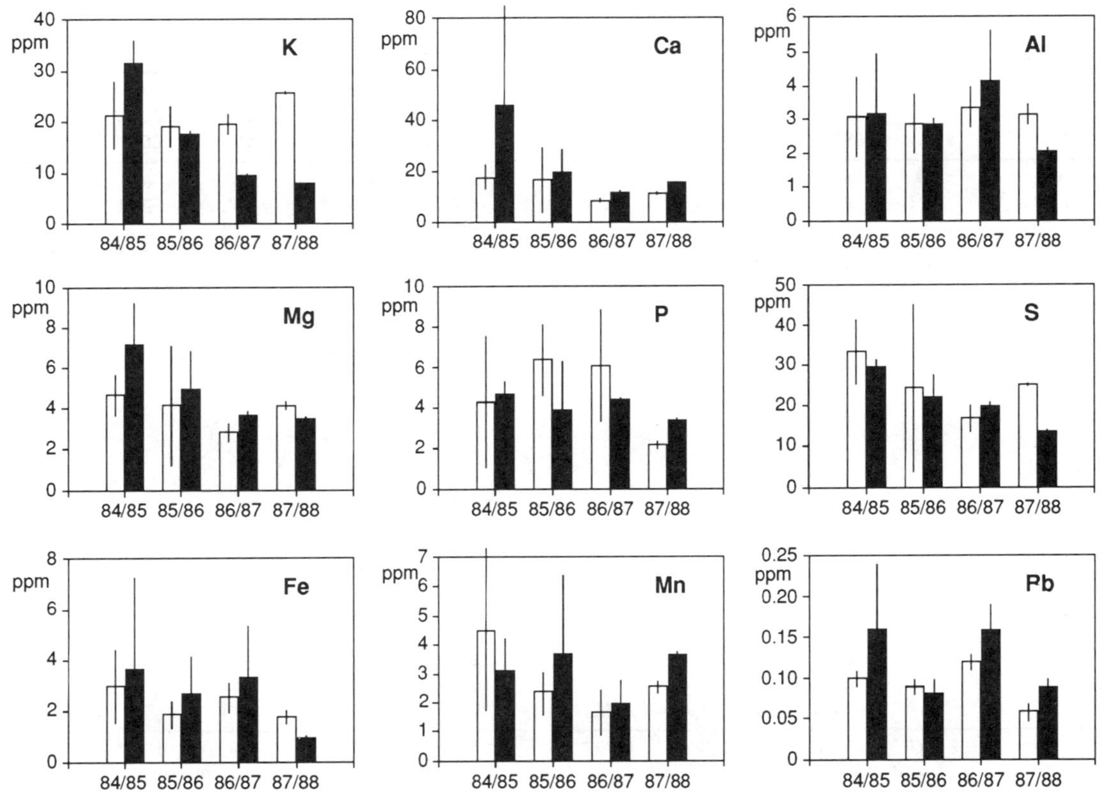

Fig. 11.2. Element content (ppm) of the soil solution in the horizon O_{fh}/A (+8 to −5 cm) during the winter periods 1984 to 1988. Soil solution obtained by centrifugation. *Open bars* Damage class 0; *full bars* damage class 3

Table 11.1. Seasonal variation of the element concentrations in the soil solutions of the horizons O_{fh}/A (+8 to −5 cm) and B_v (−5 to −25 cm). All data in mg/l. D.c. = damage class

Horizon	Element	Winter		Spring		Summer		Autumn		Mean value 1986–1989	
		D.c. 0	D.c. 3	D.c. 0	D.c. 3	D.c. 0	D.c. 3	D.c. 0	D.c. 3	D.c. 0	D.c. 3
	Mg	6.2	4.8	7.0	6.2	7.0	6.8	9.6	8.3	7.5	6.8
O_{fh}/A	Ca	20.2	18.0	19.0	24.6	23.3	24.3	31.3	26.8	23.4	23.6
	K	18.2	14.4	26.6	19.7	25.6	19.4	28.2	18.8	23.4	18.9
	Al^{3+}	1.5	1.5	2.8	1.3	3.2	2.5	2.9	2.6	2.6	2.0
	Mg	7.3	4.9	10.3	7.6	8.4	5.8	13.8	8.9	10.0	6.8
	Ca	16.3	12.3	20.8	18.6	23.3	17.4	27.8	20.3	22.1	17.2
Bv	K	11.3	7.0	14.4	9.9	16.8	10.4	15.3	11.0	14.5	9.6
	Al^{3+}	22.7	7.8	18.7	7.4	20.1	10.8	20.0	10.7	20.4	9.2

Table 11.2. Seasonal variations of the element ratios in the soil solutions of the horizons O_{fh}/A (+8 to −5 cm) and B_v (−5 to −25 cm) on the basis of data from 1986 to 1989

Horizon	Element ratio	Winter		Spring		Summer		Autumn		Mean value 1986–1989	
		D.c. 0	D.c. 3	D.c. 0	D.c. 3	D.c. 0	D.c. 3	D. c. 0	D.c. 3	D.c. 0	D.c. 3
	Mg/Al^{3+}	4.1	3.2	2.5	4.8	2.2	2.7	3.3	3.2	2.9	3.4
O_{fh}/A	Ca/Al^{3+}	13.5	12.0	6.8	18.9	7.3	9.7	10.8	10.3	9.0	11.8
	K/Ca	0.9	0.8	1.4	0.8	1.1	0.8	0.9	0.7	1.0	0.8
	Mg/Al^{3+}	0.32	0.63	0.55	1.03	0.42	0.54	0.69	0.83	0.49	0.74
B_v	Ca/Al^{3+}	0.72	1.58	1.11	2.51	1.16	1.61	1.39	1.90	1.08	1.87
	K/Ca	0.69	0.57	0.69	0.53	0.72	0.60	0.55	0.54	0.65	0.55

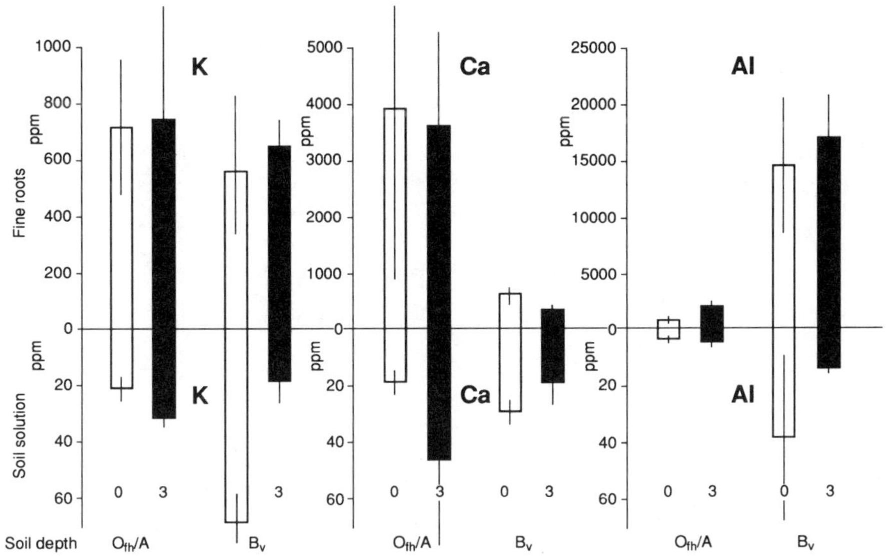

Fig. 11.3. Element content (ppm) of the soil solution and the fine roots from the root spheres of the horizons O_{fh}/A (+8 to –5 cm) and B_v (–5 to –25 cm). Data for two healthy (damage class 0) and two severely damaged trees (damage class 3) each. Sampling in February 1985

The lower part of Fig. 11.3 illustrates the pronounced increase in the Al concentration in the horizon B_v (–5 to –25 cm). A direct consequence of this is a drastic increase in the Al content in the fine roots. On the other hand, the uptake of Ca is obviously impeded. More details will be given in Section 11.2.

The seasonal variation of the element concentrations in the soil solutions of various horizons, including the humus layer, was the subject of a systematic experiment series in the period 1986 to 1989 (Rademacher et al. 1992). Again, two damage classes (0 and 3) with spruce collectives of seven trees each were taken into account. The results for some important elements are summarized in Table 11.1. The data for Al represent the Al^{3+} ion, while those for Mg, Ca and K refer to the total content in the soil solution. Along with the quarterly compiled results, the mean values for the total measuring period are also specified. Since needles of healthy trees appeared to have on the average a better supply of K and Mg than damaged collectives (Sect. 11.2), these studies were started with the hypothesis that on the very heterogeneous soil at the "Postturm" site (Ulrich 1985) healthy trees have access to a more favourable nutrient supply. This should be reflected in higher concentrations of K and Mg in the soil solution and superior K/Ca and Mg/Al^{3+} ratios, respectively. Indeed, this hypothesis is confirmed in the case of K by Tables 11.1 and 11.2. However, the conditions are much more complex for the element Mg. Table 11.1 might, at first glance, suggest that the Mg concentrations also support the above supposition, in particular

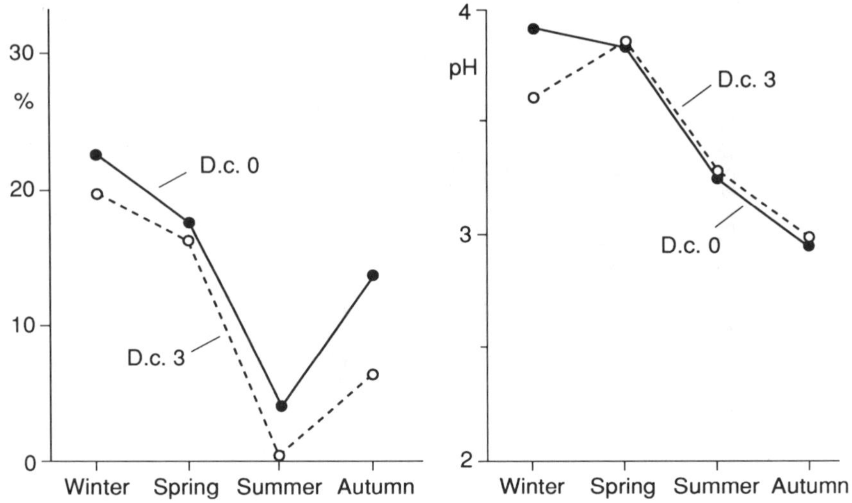

Fig. 11.4. Seasonal variations of soil moisture and pH value in the O_{fh}/A horizon of a healthy (damage class 0) and damaged spruce (damage class 3) in 1986

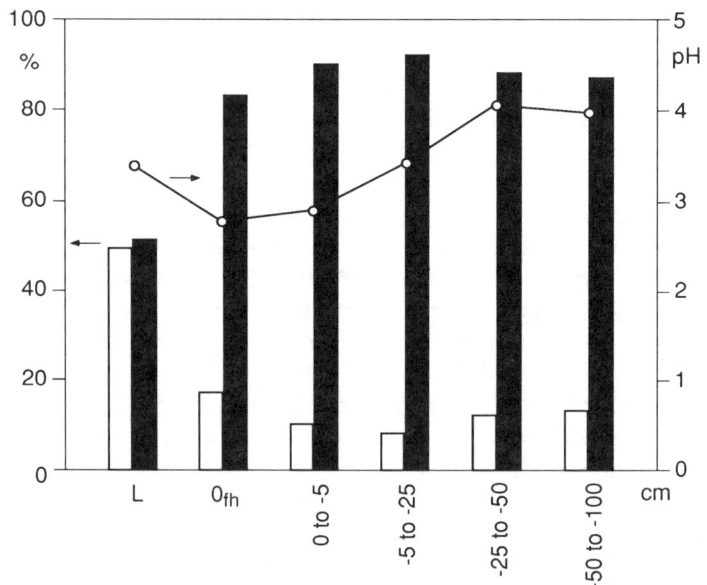

Fig. 11.5. Percentage of the total effective cation-exchange capacity for cations of strong bases (Na, Mg, K, Ca; *open bars*) and strong acids (H, Al, Mn, Fe; *full bars*) as a function of soil depth. pH values after addition of KCl

when considering the B_v horizon. However, the values for the O_{fh}/A horizon show no significant differences for the two damage classes beyond the limits of error, and the Mg/Al^{3+} ratios are mostly higher in both horizons in the case of damage class 3 (Table 11.2). The latter fact is valid for the upper stratum, above all, during the spring and summer period when active nutrient uptake occurs, and in the lower horizon throughout the whole year.

The complexity of the soil conditions also becomes obvious when the analytical data obtained in 1986 are considered. This year was characterized by an extremely low amount of precipitation at the investigation site in the period June to September. The rainfall came to only 47% of that in the same time interval in 1987. Moreover, the average air temperature was slightly higher. These meteorological conditions led to a severe decrease in the soil humidity and in the pH value of the soil solution as shown in Fig. 11.4. Analytical data for nine elements are summarized in Table 11.3. Various results clearly contradict the long-term means of Table 11.1. This refers to

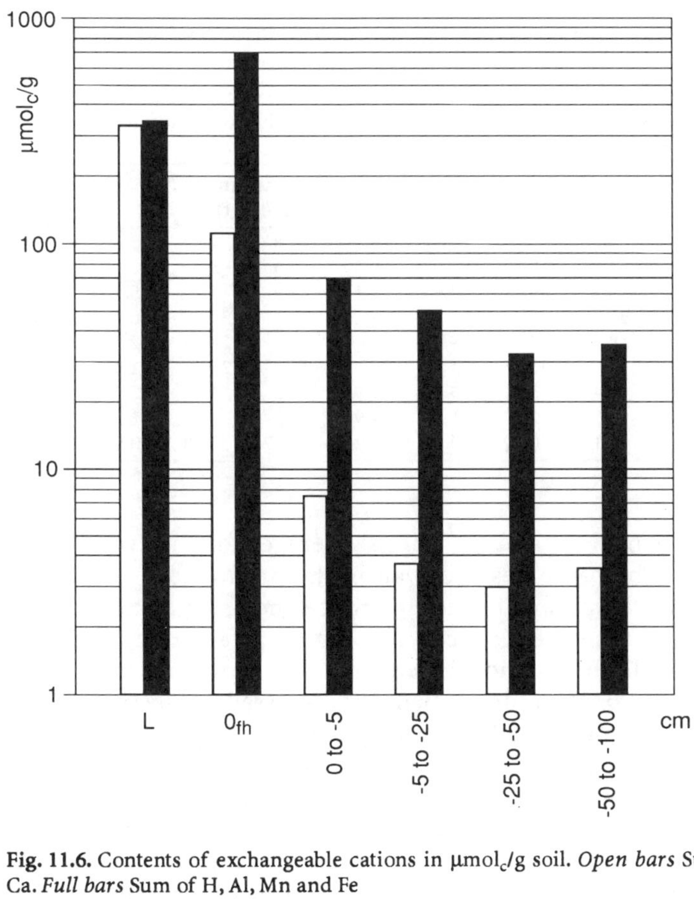

Fig. 11.6. Contents of exchangeable cations in μmol_c/g soil. *Open bars* Sum of Na, Mg, K and Ca. *Full bars* Sum of H, Al, Mn and Fe

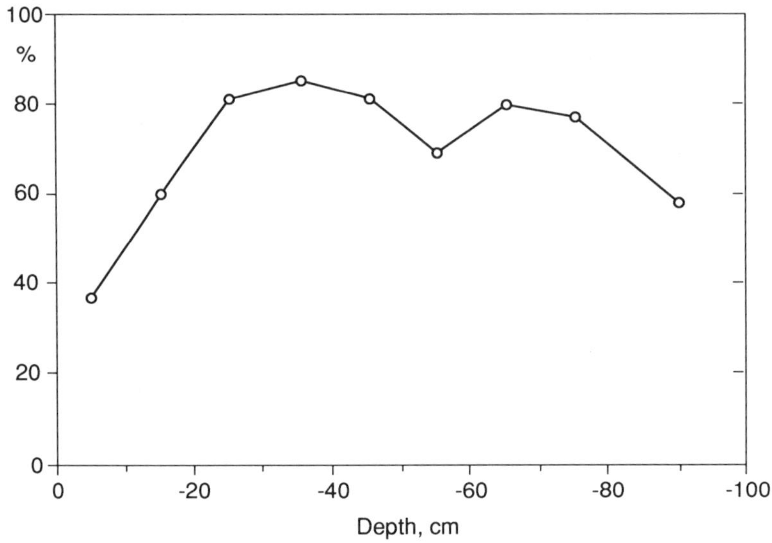

Fig. 11.7. Contribution of Al to the total cation-exchange capacity in % vs. soil depth

Table 11.3. Seasonal variation of the element concentrations in the soil solution during 1986. Horizon O_{fh}/A (+8 to −5 cm). All data in mg/l

	Winter		Spring		Summer		Autumn		Mean value	
Element	D.c. 0	D.c. 3	D.c. 0	D.c. 3	D.c. 0	D.c. 3	D.c. 0	D.c. 3	D.c. 0	D.c. 3
Mg	4.5	3.5	3.0	5.3	6.3	19.4	6.8	7.8	5.2	9.0
Al	3.3	3.9	2.8	3.4	2.9	9.3	2.5	3.4	2.9	5.0
P	7.9	5.1	5.1	5.1	3.9	9.1	5.9	4.7	5.7	6.0
S	28	20	30	37	30	88	33	33	30	45
K	22	14	18	24	19	35	28	28	22	25
Ca	18	12	14	23	24	70	26	26	21	33
Mn	2.0	1.5	1.4	3.1	3.1	9.9	3.2	4.0	2.4	4.6
Fe	2.0	3.4	2.0	2.7	1.1	2.9	0.7	1.9	1.5	2.7
Pb	0.10	0.15	0.11	0.12	0.06	0.23	0.07	0.08	0.09	0.15

both the seasonal behaviour and the annual averages for the two damage classes considered. The cause lies above all in the pronounced increase in the soil solution concentrations in the fine root spheres of damaged trees during the dry spell. These effects can amount to a factor of 3 or even greater and concern both nutritional elements and competing toxic metal ions such as Al and Pb. More details will be discussed in Section 11.2 in connection with the element uptake into the fine roots.

Table 11.1 already demonstrated the rapid decline of the nutrient supply with increasing depth in the upper soil layers and the concurrent huge rise in the concentration of Al^{3+}. These unfavourable conditions also exist in deeper strata. Some quantities characterisic for the chemical soil quality are presented in Figs. 11.5 and 11.6. The data can be regarded as being representative of the investigation site, since they are based on the average of 32 profiles. Figure 11.5 illustrates the percentage of the total effective cation-exchange capacity for cations of strong bases (Na, Mg, K, Ca) and strong acids (H, Al, Mn, Fe) as a function of soil depth. In Fig. 11.6, the corresponding content of exchangeable cations in $\mu mol_c/g$ soil is shown. In the first of these figures, the course of the pH value is also indicated. The humus layer was divided into the L (+12.5 to +8 cm) and the O_{fh} (+8 to 0 cm) horizons. The main contribution to the second group of ions has to be ascribed to Al as is shown in Fig. 11.7. These data originated from a profile taken close to the GKSS measuring tower.

The results summarized in the Tables 11.1 to 11.3 and the Figs. 11.1 to 11.7 indicate that the physiological conditions for the forest stand at the investigation site are quite poor (Foy 1974; Tischner et al. 1983; Murach 1984). The pH values in the humus layer range from 2.7 to 3.5, in the upper mineral soil (0 to -5 cm) they are about 2.9 and in the deeper horizons they range from 3.2 to about 4.0. Thus, these strata have to be assigned to the Fe, Al/Fe and Al buffer regions, respectively. On the average, the supply of the exchangeable cations Na, Mg, K and Ca in the upper 50 cm of the mineral soil amounts to at most 12% of the total exchange capacity.

The analysis shows that the nutritional elements are concentrated in the humus layer. However, a further decline in the soil fauna and other saprophytes as well as a progressive accumulation of organic necromass might sometime also reduce the availability of nutrients in this horizon (Matzner 1988).

The concentration of the nutritional elements upon the upper layers becomes likewise evident when the element stores in the different strata are regarded. Results obtained for the intensive rooting horizons L + O_{fh} (+12.5 to 0 cm) and A + B_v (0 to -50 cm) as well as the restrictive rooting horizon C_v (−50 to −100 cm) are summarized in Table 11.4. The data denote the total stores in the case of the humus layer and stand for the exchangeable stores in the case of the mineral soil. The high values in the lowest horizon (C_v) result from the large mass of this stratum and are not due to high concentrations. As has been shown for the soil solution phase, the competition of Al is

Table 11.4. Element stores (kg/ha) of the soil horizons L + O_{fh} (+ 12.5 to 0 cm), A + B_v (0 to -50 cm) and C_v (−50 to −100 cm). Definitions, see text

Horizon	Mass	Element stores (kg/ha)						
	(t/ha)	Na	Mg	Al	K	Ca	Mn	Fe
L + O_{fh}	253	29	90	631	115	528	80	1069
A + B_v	6467	122	36	1568	125	236	69	127
C_v	7805	282	56	1636	113	177	26	18
L + O_{fh} + A + B_v + C_v	14525	433	182	3835	353	941	175	1214

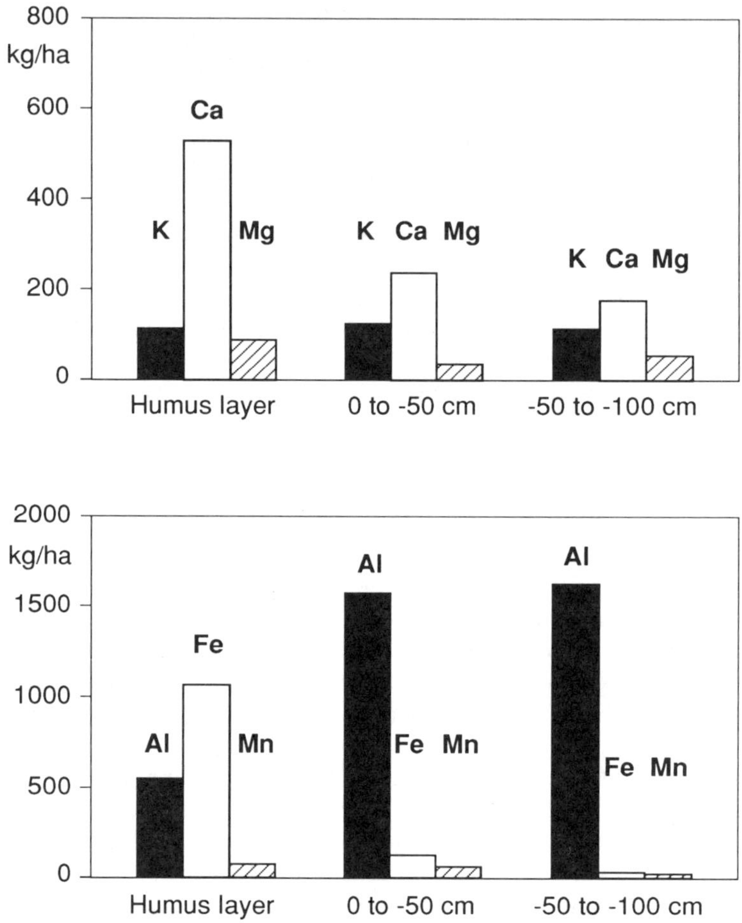

Fig. 11.8. Element stores in the humus layer and the mineral soil in kg/ha. Humus layer: total stores; mineral soil: exchangeable stores (see text)

severe so that the development of an efficient fine root system in this horizon is strongly impeded. With a Ca/Al ratio considerably below 0.5 marked damage of the fine roots and growth retardation have to be expected (Rost-Siebert 1983; Ulrich et al. 1984). It is therefore understandable that merely about 20% of the total fine root mass were found in the C_v horizon (Rademacher et al. 1992). This withdrawal of the root system towards the upper horizons is also responsible for the insufficient stability of the trees and the resulting increase in wind-induced damages. Also, the function of the soil as the essential water reservoir is utilized inadequately.

A comparison of the element stores in the mineral soil above 50 cm at the "Postturm" site with data from three other damaged northwest German forest stands (Bredemeier 1987) reveals that in the case of Mg and K impoverishment at the Postturm site has already reached a critical level. The impact by exchangeable H, Al, Mn and Fe, on the other hand, may be classified as medium-sized. In the case of Ca, the conditions are comparatively favourable. Figure 11.8 illustrates for some essential elements the stores in the humus layer as well as in the upper and lower mineral soil on the basis of the definitions given above with respect to Table 11.4.

11.2 Biomass and Element Content in the Tree Compartments

Seasonal variations in the soil moisture and pH value cause distinct alterations in the element concentrations of the soil solution. This has already been demonstrated in Fig. 11.4 and Table 11.3 on the basis of data obtained in 1986 for the O_{fh}/A horizon. Depending on the element considered and the state of health of the trees, the effects on the soil solution are passed on to the element content in the fine roots (Rademacher et al. 1988). In Figs. 11.9 to 11.11, the analytical results from 1986 are summarized for both the soil solutions (Table 11.3) and the fine roots from a healthy spruce (damage class 0) and a severely damaged spruce (class 3) in each case. Already at first sight the quite different course of the concentrations in the soil solution for the two damage classes becomes evident. In the case of the declining tree, there is a pronounced rise in the content, particularly of the elements Mg, Al, P, S, K, Ca, Mn and Pb, while the reaction in the soil of the healthy tree is rather small or even contrary.

The Mg content of the fine roots increased slightly during the dry spell independent of the grade of decline (Fig. 11.9). On the other hand, the elements Al, P, Ca, Mn, Fe and Pb exhibit a decrease in the concentration, in either both damage classes as in the case of Al or only in one of them: P, Ca and Mn in the roots of the damaged spruce, Fe and Pb in those of the healthy tree (Figs. 11.9 to 11.11). The variations of the S and K content are only moderate. Al, S, Mn and Pb were found with higher concentrations in damage class 3 throughout the year.

The supply of nutritional elements to the trees depends not only on the concentration in the soil solution, but also on a sufficient availability of soil water. In this respect, the sandy soil at the "Postturm" site with a low content of clay presents unfavourable preconditions. During dry spells, the transport into the fine roots may be strongly impeded. In 1986, the drop of the soil moisture was particularly pronounced (cf. Fig. 11.4). This may be the reason that, for instance, in the case of damage class 3, the concentrations of P and K in the fine roots decreased in spite of the increase in the soil solution (Figs. 11.9 and 11.10).

Several authors consider the Al ion as a competitive element which in the presence of low pH values displaces elements such as Mg and Ca from the anionic exchange places of the fine-root cell wall (Ulrich 1981; Bauch

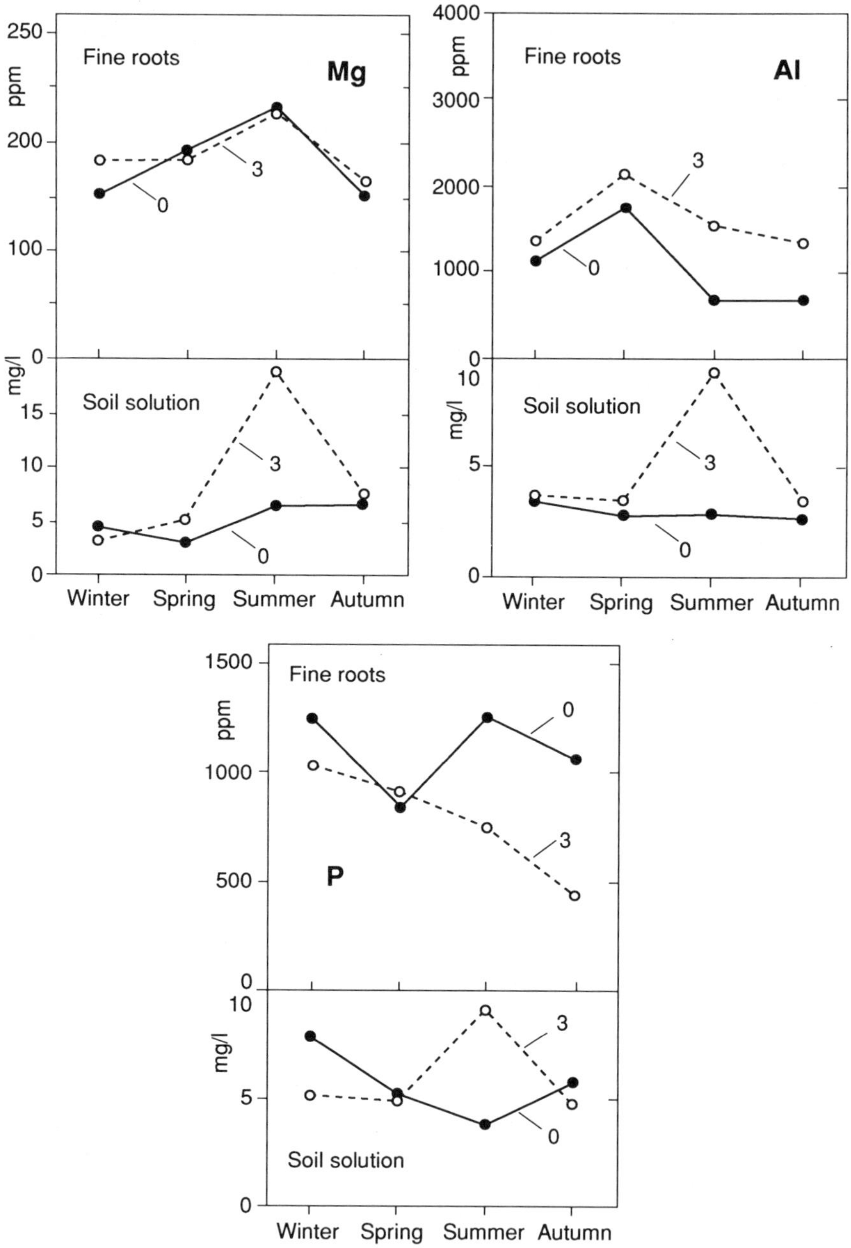

Fig. 11.9. Seasonal variations in the concentrations of Mg, Al and P in the fine roots and their surrounding soil solution from a healthy spruce (damage class 0) and a damaged spruce (class 3). Horizon O_{fh}/A

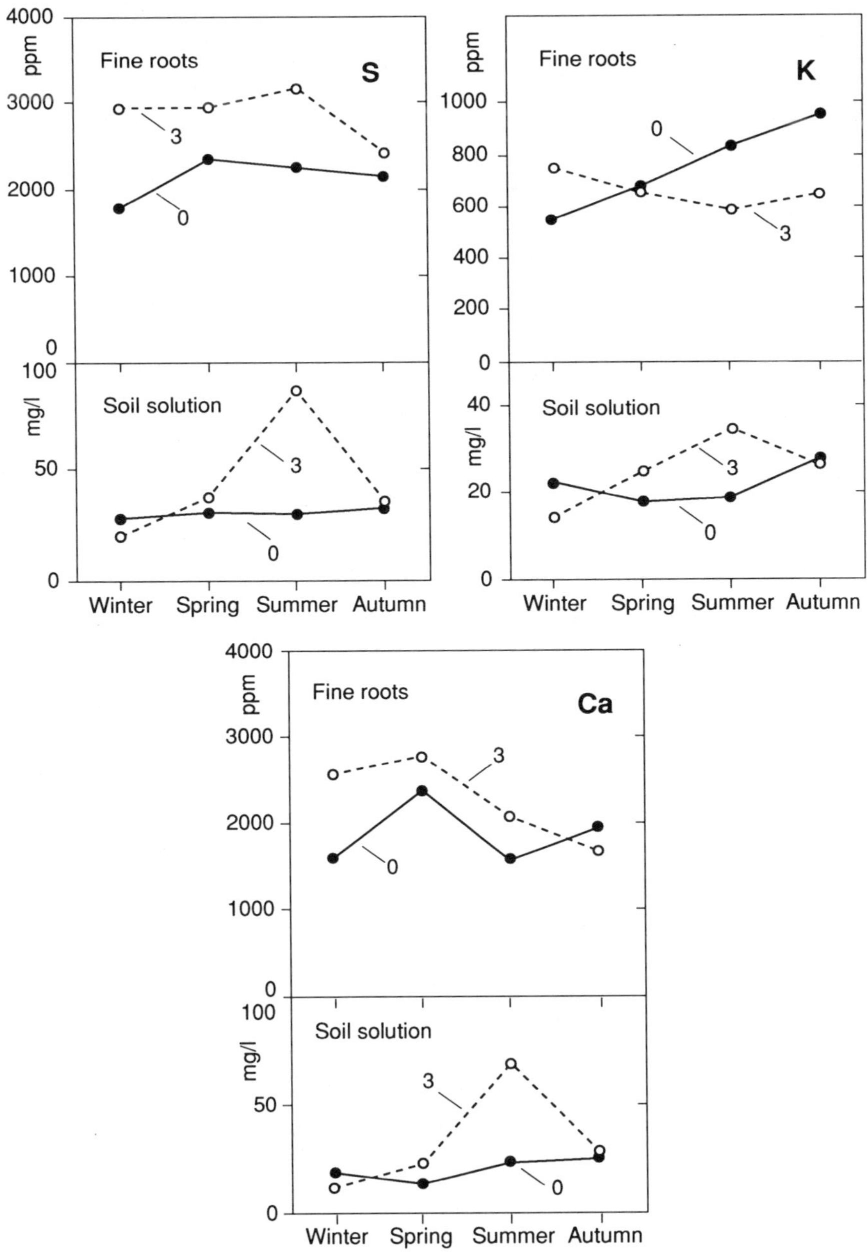

Fig. 11.10. Seasonal variations in the concentrations of S, K and Ca in the fine roots and their surrounding soil solution from a healthy spruce (damage class 0) and a damaged spruce (class 3). Horizon O_{fh}/A

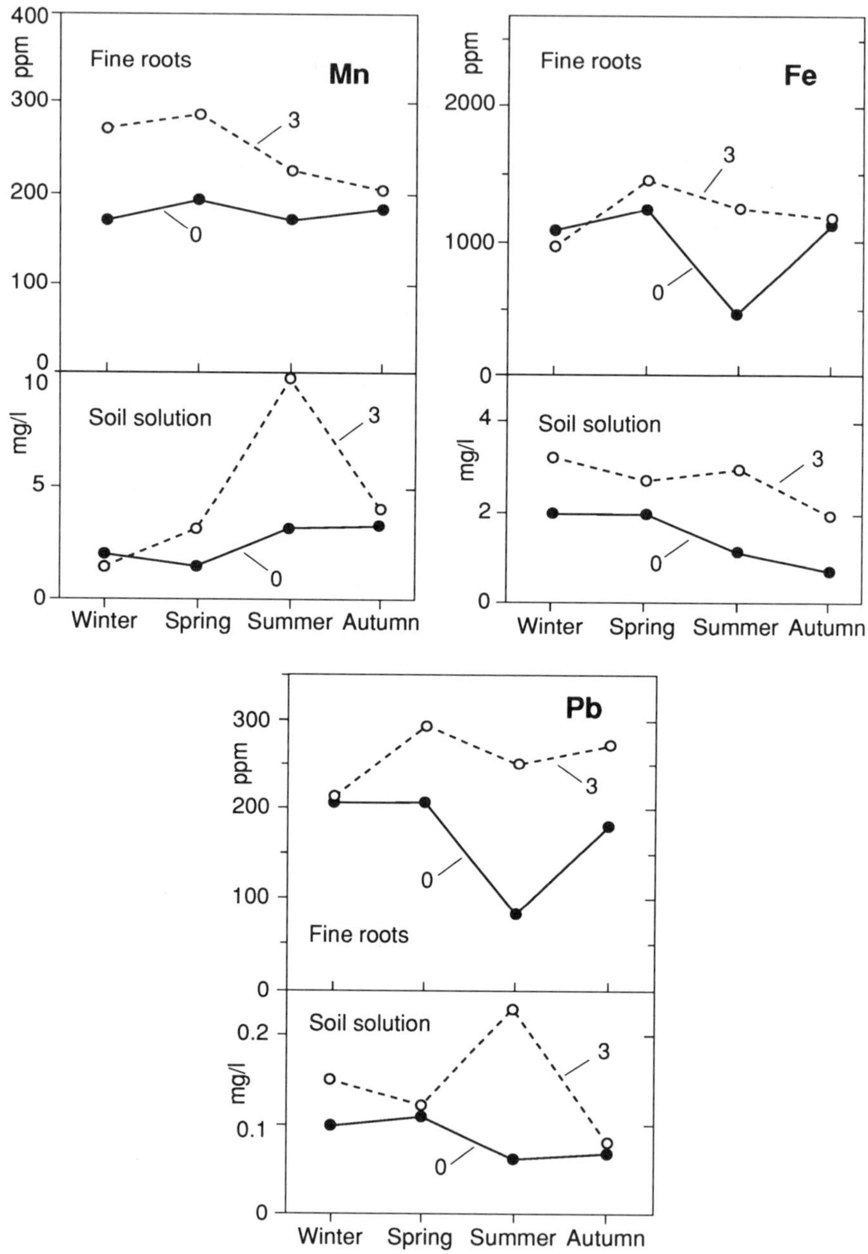

Fig. 11.11. Seasonal variations in the concentrations of Mn, Fe and Pb in the fine roots and their surrounding soil solution from a healthy spruce (damage class 0) and a damaged spruce (class 3). Horizon O_{fh}/A

and Schröder 1982; Tischner et al. 1983; Ulrich et al. 1984). This conception was corroborated by hydroponic culture tests (Stienen and Bauch 1988) and fertilization experiments with old spruce trees (Bauch et al. 1985). Obviously, these phenomena also occur at the site of the present study, in particular in the mineral soil where the competition of Al is very dominant.

In early investigations on the content of the heavy metal Pb in plants, only a minor toxical effect was ascribed to this element (Heilenz 1970). However, decades of Pb deposition together with the input of other atmospheric pollutants have evidently changed the conditions for the formerly assumed immobility. Studies in the 1980s showed that lead accumulated in the soil, as a consequence of rising acidification, is dissolved below pH 4 and that this process is enhanced by high concentrations of sulphate and chloride in the soil solution (Brümmer and Herms 1983; Herms and Brümmer 1984). Organic substances from the decomposition of litter can also raise the solubility. The present study substantiates that these factors are to a high degree valid at the "Postturm" site. The Pb concentrations in the fine roots were as high as 300 ppm (cf. Fig. 11.11). Godbold (1984) showed that the availibility of just 0.1 ppm already reduces the growth of the fine roots of spruce seedlings by more than 50%.

Nutrition physiology studies have elaborated comprehensive data on the threshold values of elements in leaves and needles (Bergmann 1983; Bosch et al. 1983; Zech and Popp 1983; Isermann 1985; Knabe et al. 1986). Last year's spruce needles should contain more than 5000 ppm K, 1000 ppm P, 1000 ppm Mg, 50 ppm Mn and 13 ppm Zn. For Ca, which usually shows deficiencies particularly in the first year, a threshold value of 3000 ppm is specified (Isermann 1985). Hüttl (1987) stated limits which were about 20% lower. However, a fixed threshold is in any case difficult to define since other factors such as different site conditions, tree age and the extent of competing elements must also be taken into account. Data obtained at the "Postturm" site for 1- to 5-year-old needles are summarized in Fig. 11.12. Again, samples from both healthy and damaged trees were analyzed. The measurement period was the winter of 1983/1984 to the winter of 1987/1988. The data represent mean values from a total of 22 trees (Rademacher et al. 1988). They reveal that the youngest needles of damage class 0 with nearly 6000 ppm were sufficiently supplied with K, while the needles of the same age, but from trees of damage class 3 already showed deficiency symptoms which further intensified with advancing years. The Ca content was approximately adequate, whereas the supply of Mg turned out to be rather poor. A lack of K is obviously accompanied by an enrichment with Ca and particularly with Pb. According to Hüttl (1987), drastic reductions in the K uptake may be expected by a Ca antagonism. Specific fertilizing tests in a southwest German spruce stand improved the K contents and the vitality of the needles in a short time. Similar success was achieved by fertilizing in the case of Mg deficiency (Zech and Popp 1983; Hüttl 1987; Isermann 1987).

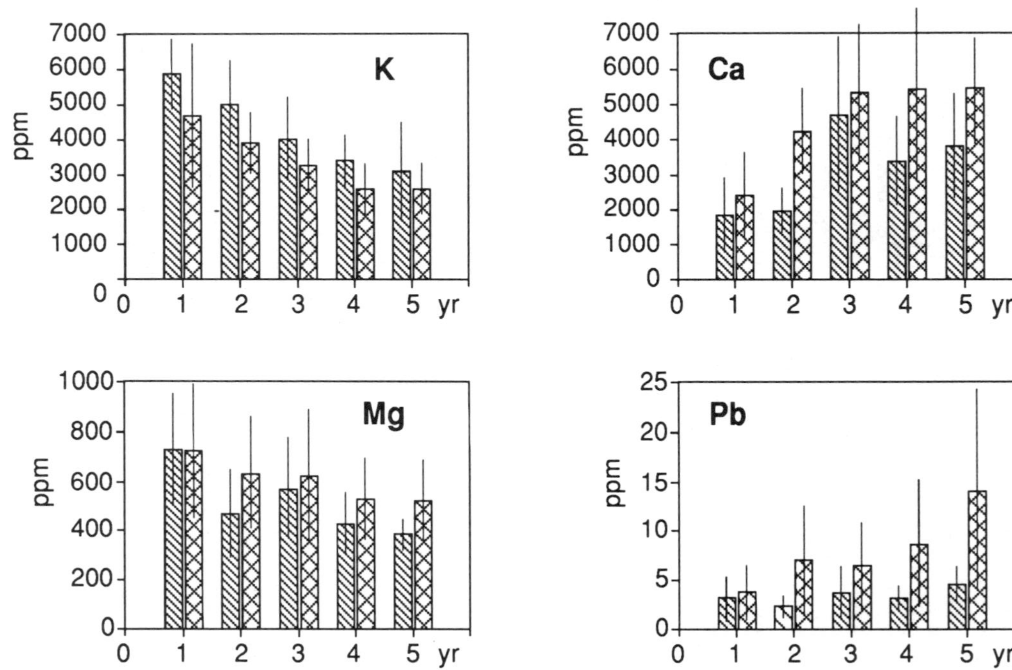

Fig. 11.12. Element content of 1- to 5-year-old needles of healthy (damage class 0; *hatched bars*) and damaged trees (damage class 3; *cross-hatched bars*). Measurements during the vegetation rests

Table 11.5. Seasonal variation of the element content (μg/g) in 3-year-old needles. Mean values for the period 1986 to 1989. D.c. = Damage class

	Winter		Spring		Summer		Autumn		Annual mean	
Element	D.c. 0	D.c. 3	D.c. 0	D.c. 3	D.c. 0	D.c. 3	D.c. 0	D.c. 3	D.c. 0	D.c. 3
Mg	641	534	583	550	588	569	640	647	613	575
Ca	4503	3433	4028	4227	4219	5374	4497	5562	4312	4649
K	4052	3463	4320	3705	4351	3596	4343	3777	4267	3635
Al	186	198	191	207	195	208	233	214	201	207

With regard to the toxical heavy metal Pb, it is important to point out the presumption that this element can also be taken up directly from the atmosphere by the needles, in particular when the cuticle is impaired. Lötschert and Grosch (1984) concluded on the basis of studies on yew-tree needles from the Frankfurt conurbation that plumbiferous airborne particulates with diameters up to 5 μm may be capable of being taken up via the stomata. Since a considerable portion of airborne particulates belongs to this size fraction (Michaelis et al. 1988, 1989, 1992a,b,c; cf. also Fig. 9.1), the principal requirements for the process are fulfilled. In glasshouse experiments, Dollard (1986) was able to follow the transport of lead taken up by radish leaves using the radioactive isotope Pb^{210} as a tracer. The uptake increased by about a factor of 5 in the case of severe damage of the cuticle. Such an impairment may, among others, be caused by NO_2 with the consequence that also washout effects, in particular of K, become a risk to the vitality of the needles.

The seasonal variation of the element content in the needles at the investigation site is illustrated in Table 11.5 (Rademacher et al. 1992). The data are mean values for the period 1986 to 1989 and refer to 3-year-old needles distinguished again between the damage classes 0 and 3. As already shown in Fig. 11.12, the Mg supply turns out to be rather poor. The K deficiency is evident throughout the year, in particular in declining trees. As expected, the Ca content of this age group is above the threshold value. In the needles of damage class 3, the concentration is altogether higher than in those of the healthy trees. This phenomenon is possibly due to the higher consumption of water in the presence of enhanced K deficiency and the associated irreversible enrichment of Ca in the needles (Linser and Herwig 1968; Bangerth 1979).

Compared to the declining trees at the "Postturm" site, the healthy spruces are characterized by stronger growth and greater biomass (Bauch et al. 1986; Rademacher 1986; Bauch et al. 1988). Under this aspect and considering also the higher Mg content in these trees, an increasing deficiency of Mg in the surroundings of their fine roots is to be expected in the poor soil (Rademacher et al. 1992). Such a trend will be accompanied by an enhanced impact of Al^{3+}. Processes of this kind have been studied by Huikari (1977).

Table 11.6. Element contents of the tree components in ppm. Mean values from up to 30 tree inventories

	Element content (ppm)						
Compartment	Na	Mg	Al	K	Ca	Mn	Fe
Needles/1 year	640	760	130	5720	2550	1210	160
Needles/2 year	870	700	175	4790	4220	1950	200
Needles/3 year	1340	610	210	3950	5520	2640	190
Needles/4 year	1230	520	220	3620	5830	2770	190
Needles/5 year	1180	500	260	3180	6060	2800	220
Needles/> 5 year	1180	500	260	3180	6060	2800	220
Branch/bark	660	730	650	3660	5180	1170	1120
Branch/wood	330	320	41	2290	1530	625	42
Bough/bark	660	900	200	2520	10900	1890	210
Bough/wood	20	200	8	830	1310	420	4.5
Stem/bark	100	140	130	330	7530	900	140
Stem/bast	130	650	56	3080	8410	1930	41
Stem/sapwood	45	82	8	650	550	200	3.4
Stem/heartwood	50	130	9	490	705	230	1.4
Root stock/bark	130	650	56	3080	8410	1930	41
Root stock/sapwood	54	82	8	650	550	200	3.4
Root stock/heartwood	51	130	9	490	705	230	1.4
Main root/O_{fh} A/bark	220	560	125	2540	6920	1110	50
Main root/O_{fh} A/wood	160	300	19	1980	1360	510	8.8
Main root/B_v I/bark	750	610	940	2120	6420	1090	76
Main root/B_v I/wood	700	290	26	1610	820	380	9
Main root/B_v II/bark	280	360	1530	1520	3450	540	140
Main root/B_v II/wood	150	230	24	1850	760	310	13
Main root/C_v/bark	280	360	1530	1520	3450	540	140
Main root/C_v/wood	150	230	24	1850	760	310	13
Fine root/O_{fh} A	6400	270	1880	890	2540	270	1820
Fine root/B_v I	5820	450	12900	1110	1200	220	4920
Fine root/B_v II	12700	550	17800	2010	1210	320	6470
Fine root/C_v	12700	550	17800	2010	1210	320	6470

A useful support for a differentiated evaluation of the supply, the inventory and the losses of elements is provided by a detailed analysis of the element content, the biomass and the storage in the tree compartments. The extensive data sets collected at the "Postturm" site from up to 30 spruce trees are summarized in the form of mean values in Tables 11.6 and 11.7 (Rademacher et al. 1992). Details of the biomass determination are given elsewhere (Rademacher 1986; Rademacher et al. 1986; Murach et al. 1988; Dünisch et al. 1992). Certainly, the data given in the tables are subject to possible errors by virtue of the necessary projection, they are, however, considered as a sufficient basis for the quantification and evaluation of the element inventories. A tree population of 450 per ha was used for the projection.

Table 11.6 shows that the elements Mg, Al, K and Ca exhibit quite different distribution patterns. Mg reveals a rather uniform distribution of the content with the exception of the sapwood and heartwood. On the other

Table 11.7. Biomass and element stores of the tree components in t/ha and kg/ha, respectively. Mean values from up to 30 tree inventories

	Biomass	Element stores (kg/ha)						
Component	(t/ha)	Na	Mg	Al	K	Ca	Mn	Fe
Needles/1 year	5.4	3.4	4.1	0.69	30.6	13.7	6.5	0.85
Needles/2 year	5.2	4.5	6.6	0.91	24.8	21.9	10.1	1.02
Needles/3 yr	5.0	6.6	3.1	1.02	19.6	27.3	13.1	0.96
Needles/4 year	4.2	5.1	2.2	0.93	15.1	24.4	11.6	0.80
Needles/5 year	2.7	3.2	1.4	0.71	8.6	16.4	7.5	0.60
Needles/> 5 year	1.0	1.2	0.52	0.27	3.3	6.3	2.9	0.23
Branch/bark	6.2	4.1	4.6	4.05	22.7	32.1	7.3	6.94
Branch/wood	4.1	1.4	1.3	0.17	9.5	6.3	2.6	0.17
Bough/bark	19.7	13.0	17.6	4.01	49.6	214.0	37.2	4.13
Bough/wood	39.3	0.79	8.0	0.30	32.8	51.7	16.6	0.18
Stem/bark	9.1	0.91	1.2	1.16	3.0	68.8	8.3	1.28
Stem/bast	13.6	1.8	8.8	0.76	41.7	113.9	26.1	0.55
Stem/sapwood	246.6	11.1	20.1	2.00	159.6	136.4	49.1	0.84
Stem/heartwood	131.4	6.7	16.6	1.13	64.7	92.6	30.8	0.18
Root stock/bark	2.3	0.30	1.5	0.13	7.1	19.3	4.4	0.09
Root stock/sapwood	23.7	1.1	1.9	0.19	15.3	13.1	4.7	0.08
Root stock/heartwood	12.7	0.64	1.6	0.11	6.2	8.9	3.0	0.02
Main root/O_{fh} A/bark	5.5	1.2	3.0	0.68	13.9	37.7	6.0	0.27
Main root/O_{fh} A/wood	21.9	3.4	6.5	0.42	43.3	29.8	11.2	0.19
Main root/B_v I/bark	4.1	3.1	2.5	3.84	8.7	26.3	4.5	0.31
Main root/B_v I/wood	16.4	11.4	4.8	0.42	26.3	13.5	6.3	0.15
Main root/B_v II/bark	2.9	0.83	1.1	4.48	4.5	10.1	1.6	0.42
Main root/B_v II/wood	11.7	1.7	2.7	0.28	21.6	8.9	3.6	0.15
Main root/C_v/bark	3.2	0.91	1.2	4.96	4.9	11.2	1.8	0.46
Main root/C_v/wood	12.9	1.9	2.9	0.30	23.8	9.8	3.9	0.17
Fine root/O_{fh} A	1.12	7.1	0.30	2.10	1.0	2.8	0.30	2.03
Fine root/B_v I	0.84	4.9	0.38	10.81	0.93	1.0	0.18	4.11
Fine root/B_v II	0.59	7.6	0.32	10.59	1.2	0.72	0.19	3.84
Fine root/C_v	0.66	8.4	0.36	11.71	1.3	0.79	0.21	4.25
Total	614	118	127	69	666	1020	282	35

hand, the elements Na, Al and Fe are strongly enriched in the fine roots with a trend increasing towards deeper horizons. Ca shows extremely high concentrations in the bark and bast components, the lowest values occur in the sapwood and heartwood. The element K has the highest content in the needles showing a decreasing tendency with age.

Multiplication of the element contents of Table 11.6 by the respective biomass (dry weight) reveals the element stores in the various components as compiled in Table 11.7. A summary of this detailed schedule is given in Table 11.8. The data demonstrate that the total stores of Mg in the biomass with about 120 kg/ha already exceed the total exchangeable stores in the mineral soil which amount to less than 100 kg/ha (cf. Table 11.4). Only if the stores bonded in the humus layer (90 kg/ha) are also included in these considerations, is the reservoir in the soil moderately higher than the amount of Mg fixed in the live biomass. These results may have grave consequences for

Table 11.8. Distribution of biomass and element stores over the main tree compartments on the basis of Table 11.7

Compartment	Biomass (t/ha)	Element stores (kg/ha) Na	Mg	Al	K	Ca	Mn	Fe
Needles	23.5	24.0	17.9	4.5	102	110	51.7	4.5
Branches and boughs	69.3	19.3	31.5	8.5	115	304	63.7	11.4
Stem/bark and bast	22.7	2.7	10.0	1.9	45	183	34.4	1.8
Stem wood	378	17.8	36.7	3.1	224	229	79.9	1.0
Stock and main root	117	26.5	29.7	15.8	176	189	51.0	2.3
Fine roots	3.2	28.0	1.4	35.2	4.4	5.3	0.9	14.2
Total	614	118	127	69	666	1020	282	35

the timber industry. It is true that, in the case of systematic utilization, not the total biomass of a tree with the corresponding Mg content is taken from the ecosystem, but already the withdrawal of the stem wood from the forest stand with a loss of more than 35 kg/ha would be equivalent to the total amount of exchangeable Mg in the intensive rooting horizon (A + B_v; 0 to – 50 cm). This means that under unchanged environmental boundary conditions already the next tree generation has to grow with a critical Mg supply. Deeper horizons cannot provide compensation by virtue of the pronounced impact of Al^{3+} ions. The flat rooting system of a young spruce stand is in any case precluded from such a supply. The problem of complete inclusion of the humus layer regarding these aspects has already been discussed in Section 11.1.

Even more severe is the stocktaking in the case of K. Though the exchangeable stores in the (0 to –50 cm) horizon of the mineral soil with about 125 kg/ha are markedly higher than those of Mg (Table 11.4), this has to be seen in relation to the demand which alone for the stem wood of a full-grown spruce stand amounts to considerably more than 200 kg/ha. In the case of Mn, no risks have to be expected in the near future. This conclusion is based on two factors: (1) the stores determined in the soil and in the trees as well as (2) the assumption that this element is usually taken up by the biomass proportional to the available supply without severe nutrition-physiological effects. As concerns the element Ca, the data in Tables 11.4, 11.7 and 11.8 do also not indicate a significant deficiency situation in the immediate future. The available stores in the upper 50 cm of the mineral soil clearly exceed the demand, at least in the case of the stem wood. However, in all these considerations, the balances of the element fluxes described in Section 11.3 must also be included. In particular, in the case of Mg and Ca, there is a yearly net export via the seepage water in the order of 10 kg/ha × year. The atmospheric deposition cannot compensate for the losses from the ecosystem. The silicate decomposition rate which may be assumed to amount to 0.02 to 0.10 $kmol_c$/ha × year for 1 m soil depth does not appreciably improve these balances. If the shares of Ca, K and Mg are about 60, 25 and 15%, respectively, this would be equivalent to positive effects of only 0.04 to 0.18 kg Mg/ha × year and 0.2 to 1.0 kg K/ha × year. In the case of Mg, the result amounts to just 1% of the yearly deficit. The maximum gain of K corresponds to no more than 50% of the annual demand of the stem wood only.

11.3 Balance of Element Fluxes

The quantification of the water fluxes is one of the most important preconditions for the determination of the element fluxes and their balance in the forest ecosystem. The Ratzeburg area belongs to the regions in central Europe with rather low rainfall. In the period 1989 to 1991, the mean annual amount of precipitation at the "Postturm" site was 650 mm (cf. Fig. 7.7). Only about 50% of this amount penetrates the crown compartment and reaches, as the so-called throughfall, the soil surface. The rest is retained by the needles and evaporates for the most part. More than 40% of the water reaching the soil, i.e. about 140 l/m^2 × year, is taken up by the fine roots in the upper soil horizons (>–5 cm). Due to the high water conductance of the large-pored sandy soil, approximately 100 l/m^2 × year leave the rooting horizons and thus the ecosystem via the seepage water. Therefore on the average the amount of water available for transpiration is limited to slightly over 200 l/m^2 × year. The conditions found at the investigation site are summarized in Fig. 11.13 (Rademacher et al. 1992). Since below –120 cm there is no sink other than the seepage, the fluxes below this level may be set equal to

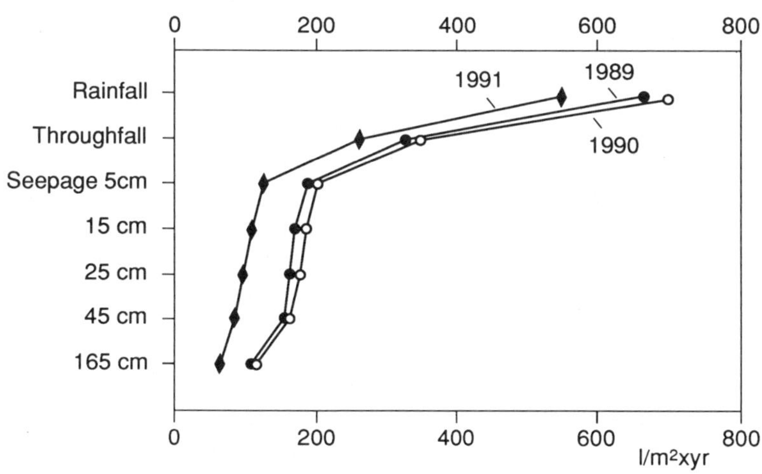

Fig. 11.13. Fluxes of water at seven measuring planes of the forest ecosystem (l/m^2 yr). Mean values for the period 1989 to 1991

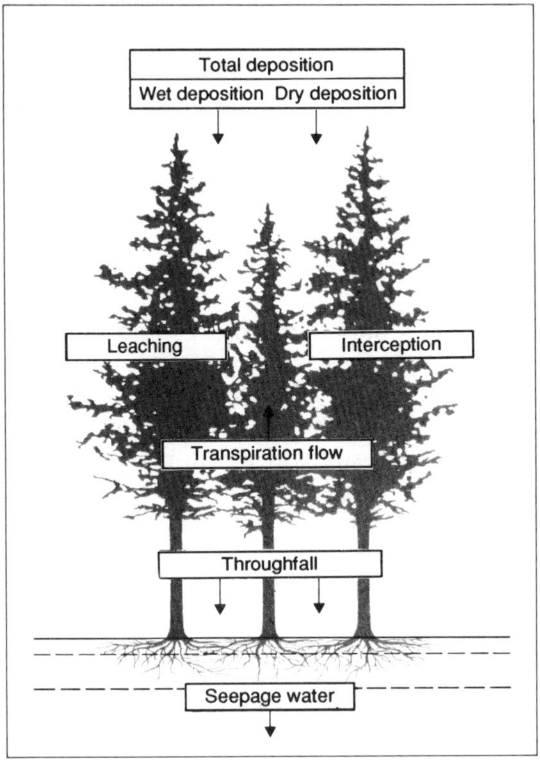

Fig. 11.14. Scheme of the element fluxes in a spruce forest ecosystem

that at −165 cm. A general remark should be made regarding the dimension of the data presented in this section. While the wet and dry deposition from the atmosphere can be determined on a short-term basis (Chaps. 7 to 10), lysimeter measurements in the soil, due to the long transport times involved, only make sense if performed over longer periods. All fluxes are therefore given on a yearly basis so that balances may also be derived.

The relevant fluxes for the element transport in the forest ecosystem are schematically displayed in Fig. 11.14. The total atmospheric deposition follows from the data obtained with the aid of wet-only samplers and multistage impactors for the wet and dry deposition, respectively (Chaps. 7 to 10). Elements are washed out from needles or bark in the canopy by acid rain; on the other hand, airborne particulates are collected in the crown compartment by interception. The throughfall contains the products of the leaching process as well as those parts of the interception which have not entered into a permanent interaction with the crown compartment. In the case of a spruce stand the throughfall is equal to the canopy drip since the stem flow is very small and therefore negligible.

Table 11.9. Water and element fluxes as well as their balances at the investigation site "Postturm". Mean values from the period 1989 to 1991

Flux	Balance	H_2O	Element fluxes ($kg\ ha^{-1} year^{-1}$)							
		($l\ m^{-2} year^{-1}$)	Mg	Al	K	Ca	Mn	Fe	Pb	H^+ion[a]
Total deposition		637	5.6	3.5	3.4	9.3	0.15	3.5	0.34	0.40
Throughfall		311	8.4	1.1	22.7	15.9	1.8	1.0	0.08	0.32
	Leaching	- 326	2.8	- 2.4	19.3	6.6	1.6	- 2.5	-0.26	- 0.08
Seepage/ 5 cm		169	10.2	10.5	12.8	28.4	6.7	3.3	0.037	-
Seepage/ 45 cm		134	8.8	20.4	2.8	12.5	3.1	0.3	0.007	0.14
Seepage/ 165 cm		98	16.5	20.8	1.0	26.3	3.0	0.2	0.007	0.08
	Total deposition minus seepage/ 45 cm	503	- 3.2	- 16.9	0.6	- 3.2	-3.0	3.2	0.33	0.26
	Total deposition minus seepage/ 165 cm	539	- 10.9	- 17.3	2.4	- 17.0	- 2.9	3.3	0.33	0.32

[a] Precipitation only

Results obtained for several relevant elements are summarized in Table 11.9 (Panten 1990; Schultz 1991; Michaelis et al. 1992a-c; Rademacher et al. 1992). All the fluxes in this schedule were determined independently. They refer to the period 1989 to 1991. The comparison of total deposition and throughfall reveals three groups of elements. To the first one belong those elements which are retained in the crown compartment. Typical examples are the heavy metals Zn and Pb. Rainwater and hydrogen ions are also subject to this retention mechanism. The second group comprises elements which in the long-term mean values predominate in the throughfall. They are obviously released from the canopy and must be restored via the roots. The most important exponents are the nutritional elements Mg, K, Ca and Mn. Finally, there is a group of elements which show a neutral behaviour. A negative sign of the leaching in Table 11.9 means enrichment in the crown compartment, a positive sign implies washout from the needles and bark. In the case of K the throughfall exceeds the atmospheric input by more than a factor of 6 so that this element takes over the main part of the proton buffering in the canopy. This process leads to a reduction of the acid impact below the crown compartment, in particular during the vegetation period, as is shown in Fig. 11.15 by means of the seasonal variation of the

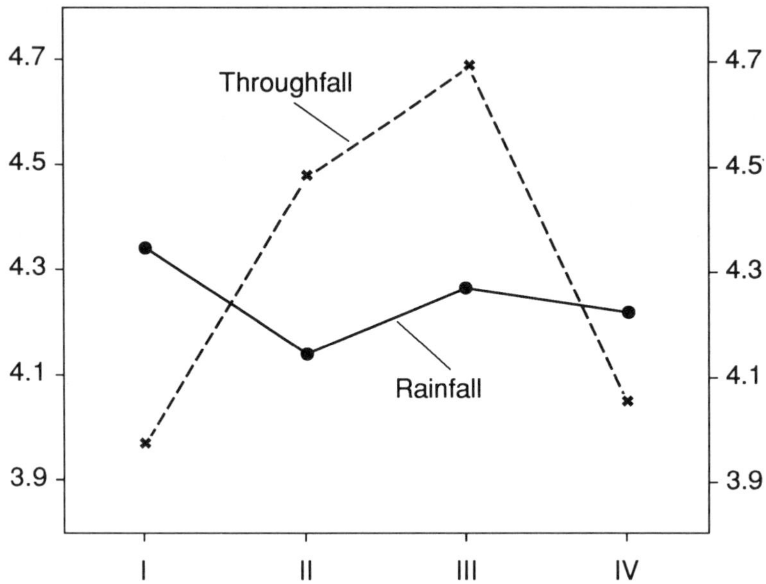

Fig. 11.15. Quarterly means of the pH values of rainfall and throughfall in 1989. *I* Jan.–March; *II* April–June; *III* July–Sept.; *IV* Oct.–Dec.

pH values in the rain and the throughfall. In contrast, according to Fig. 11.15, lower pH values may also occur underneath the tree-top, above all during the vegetation rest. Several possible causes for these findings are discussed in the literature: meteorological factors such as amount of precipitation and transmittance of the crown compartment, fog and snow weather conditions as well as acid enrichment effects by the interception process (Ulrich 1983; Adam et al. 1987; Panten 1990; Rademacher et al. 1992). The protons buffered in the crown are finally transmitted to the soil either during the restoration of the nutritional elements or with the needle fall. By the latter process also elements retained in the canopy are ultimately deposited on the soil.

As a consequence of the increasing acidification of the soil, the cations leached from the crown compartment cannot sufficiently be held back so that essential nutrients migrate via the seepage into horizons which then cannot be reached by the fine roots. Thereby these substances are lost from the ecosystem and no longer contribute to the nutrition of the plant organisms. In the final stage of this development, a deficient element supply to the forest stand occurs. The trees show characteristic damage symptoms and are no longer capable of sufficient crown buffering. At present, the removal of K is still compensated by a factor of 3 higher atmospheric deposition, though the relation to the stores in the biomass and the yearly incorporation must also be considered (Sect. 11.2). In the case of Mg and Ca, the balance turns out to be much worse. The export from the ecosystem is approximately a factor of 3 higher than the input. An extremely unfavourable ratio was found for Mn.

Toxic heavy metals such as Cd or Pb further increase these harmful impacts. Pb, for instance, deposited for many years with the needle fall onto the soil, has been fixed for decades in a physiologically ineffective form in the humus layer. With increasing acidification at pH values of about 3.2, a mobilization and availability to plants took place with the consequence of elevated Pb concentrations in the soil solution and the fine roots (Sects. 11.1 and 11.2) as well as increasing the danger of infiltration into the groundwater.

References

Adam K, Evers FH, Littek Th (1987) Ergebnisse niederschlagsanalytischer Untersuchungen in südwestdeutschen Wald-Ökosystemen 1981–1986. Kernforschungszentrum Karlsruhe, KfK-PEF 24,119 pp

Athari S, Kramer H (1983) Erfassen des Holzzuwachses als Bioindikator beim Fichtensterben. Allg Forstz 38:767–769

Bangerth F (1979) Calcium-related physiological disorders of plants. Annu Rev Phytopathol 17:97–122

Bauch J, Schröder W (1982) Zellulärer Nachweis einiger Elemente in den Feinwurzeln gesunder und erkrankter Tannen (*Abies alba* Mill.) und Fichten (*Picea abies* [L.] Karst). Forstwiss Centralbl 101:285–294

Bauch J, Klein P, Frühwald A, Brill H (1979) Alterations of wood characteristics in *Abies alba* Mill. due to "fir-dying" and considerations concerning its origin. Eur J For Pathol 9:321–331

Bauch J, Stienen H, Ulrich B, Matzner E (1985) Einfluß einer Kalkung bzw. Düngung auf den Elementgehalt in Feinwurzeln und das Dickenwachstum von Fichten aus Waldschadensgebieten. Allg Forstz 40:1148–1150

Bauch J, Göttsche-Kühn H, Rademacher P (1986) Anatomische Untersuchungen am Holz von gesunden und kranken Bäumen aus Waldschadensgebieten. Holzforschung 40:281–288

Bauch J, Göttsche-Kühn H, Riehl G (1988) Zuwachsverlust bei Fichte auf der Waldschadensfläche "Postturm" im Forstamt Farchau/Ratzeburg. In: Bauch J, Michaelis W (eds) Das Forschungsprogramm Waldschäden am Standort "Postturm", Forstamt Farchau/Ratzeburg. GKSS Forschungszentrum Geesthacht, GKSS 88/E/55, pp 289–304

Berchtold R, Alcubilla M, Evers FH, Rehfuess KE (1981) Standortkundliche Studien zum Tannensterben. Nadel- und bastanalytischer Vergleich zwischen befallenen und gesunden Bäumen. Forstwiss Centralbl 100:236–253

Bergmann W (1983) Ernährungsstörungen bei Kulturpflanzen. VEB Gustav-Fischer-Verlag, Jena, 313 pp

Bosch C, Pfannkuch E, Baum U, Rehfuess KE (1983) Über die Erkrankung der Fichten (*Picea abies* Karst.) in den Hochlagen des Bayrischen Waldes. Forstwiss Centralbl 102:167–181

Bredemeier M (1987) Stoffbilanzen, interne Protonenproduktion und Gesamtsäurebelastung des Bodens in verschiedenen Waldökosystemen Norddeutschlands. Ber des Forschungszentrums Waldökosysteme/Waldsterben, Univ Göttingen, Bd 33, 183 pp

Brümmer G, Herms U (1983) Influence of soil reaction and organic matter on the solubility of heavy metals in soils. In: Ulrich B, Pankrath J (eds) Effects of accumulation of air pollutants in forest ecosystems. D Reidel, Dordrecht, pp 233–243

Dollard GJ (1986) Glasshouse experiments on the uptake of foliar applied lead. Environ Pollut, Ser A 40:109–121

Dünisch O, Bauch J, Rademacher P, Puls J (1992) Beurteilung der Düngung eines umweltbelasteten Fichtenaltbestandes im Hinblick auf seine Stabilisierung. In: Michaelis W, Bauch J (eds) Luftverunreinigungen und Waldschäden am Standort "Postturm", Forstamt Farchau/Ratzeburg. GKSS Forschungszentrum Geesthacht, GKSS 92/E/100, pp 251–286

Elstner EF, Osswald W (1984) Fichtensterben in Reinluftgebieten: Strukturresistenzverlust. Naturwiss Rundsch 37:52–61

Eschrich W (1988) Beispiele der pflanzenphysiologischen Forschung an Bäumen im Zusammenhang mit Waldschäden in der Bundesrepublik. In: Projekt Europäisches Forschungszentrum für Maßnahmen zur Luftreinhaltung (ed) 4. Statuskolloquium des PEF. Kernforschungszentrum Karlsruhe, KfK-PEF 35, pp 25–26

Fedderau-Himme B, Feig R, Herms A, Rosengärtner K, Grothey V, Hüttermann A (1984) Untersuchung von physiologischen Parametern von Altfichten der Standorte Abt. 109 (Kammlage) und 79 (Muldenlage) im Hils. Ein Beitrag zur Quantifizierung von Hypothesen zur Ursache des Waldsterbens. Ber des Forschungszentrums Waldökosysteme/Waldsterben, Univ Göttingen, Bd 4, 127 pp

Foy CD (1974) Effects of aluminum on plant growth. In: Carson EW (ed) The plant root and its environment. Univ Press of Virginia, Charlottesville, pp 601–642

Frenzel B, Christmann A (1986) Untersuchungen zum Hormonhaushalt gesunder und walderkrankter Nadelbäume. In: Projekt Europäisches Forschungszentrum für Maßnahmen zur Luftreinhaltung (ed) 2. Statuskolloquium des PEF. Kernforschungszentrum Karlsruhe, KfK-PEF 4, pp 65–77

Gehrmann J, Gerriets M, Puhe J, Ulrich B (1984) Untersuchungen an Boden, Wurzeln, Nadeln und erste Ergebnisse von Depositionsmessungen im Hils. In: Ber des Forschungszentrums Waldökosysteme/Waldsterben, Univ Göttingen, Bd 2, pp 169–207

Godbold DL (1984) The uptake and toxicity of heavy metals in *Picea abies* (Karst.) seedlings. In: Ber des Forschungszentrums Waldökosysteme/Waldsterben, Univ Göttingen, Bd 4, pp 197–212

Hartmann P, Scheitler M, Fischer R (1989) Soil fauna comparisons in healthy and declining Norway spruce stands. In: Schulze ED, Lange OL, Oren R (eds) Forest decline and air pollution. A study of spruce (*Picea abies*) on acid soils. Ecological Studies 77. Springer, Berlin Heidelberg New York, pp 137–150

Heilenz S (1970) Untersuchungen über den Bleigehalt von Pflanzen an verkehrsreichen Standorten. Landwirtsch Forsch, Sonderheft 25 I:73–78

Herms U, Brümmer G (1984) Einflußgrößen der Schwermetall-Löslichkeit und -Bindung in Böden. Z Pflanzenernähr Bodenk 147:400–424

Hildebrand EE (1985) Ionengleichgewicht in Mineralböden von Fichtenstandorten mit nadelanalytisch festgestellten Ca^-, Mg^- und K^- Mangelzuständen. In: Projekt Europäisches Forschungszentrum für Maßnahmen zur Luftreinhaltung (ed) 1. Statutskolloquium des PEF. Kernforschungszentrum Karlsruhe, KfK-PEF 2, pp 215–227

Horn R, Zech W (1987) Zusammenhänge zwischen Bodeneigenschaften und Waldschäden. Allg Forstz 42:300–302

Horn R, Schulze ED, Hantschel R (1989) Nutrient balance and element cycling in healthy and declining Norway spruce stands. In: Schulze ED, Lange OL, Oren R (eds) Forest decline and air pollution. A study of spruce (*Picea abies*) on acid soils. Ecological Studies 77. Springer, Berlin Heidelberg New York, pp 444–455

Huikari O (1977) Micro-nutrient deficiency causes growth-disturbances in trees. Silva Fenica 11/3:251–254

Hüttl RF (1987) Neuartige Waldschäden, Ernährungsstörungen und Düngung. Allg Forstz 42:289–299

Isermann K (1985) Diagnose und Therapie der "neuartigen Waldschäden" aus der Sicht der Waldernährung. In: VDI-Berichte 560, Waldschäden – Einflußfaktoren und ihre Bewertung. VDI Verlag, Düsseldorf, pp 897–920

Isermann K (1987) Revitalisierung geschädigter Fichten-Altbestände durch Mineraldüngung. Allg Forstz 42:1–4

Jüttner F (1985) Analyse stofflicher Veränderungen in Laub und Nadelblättern immissionsgeschädigter Waldbäume. In: Projekt Europäisches Forschungszentrum für Maßnahmen zur Luftreinhaltung (ed) 1. Statuskolloquium des PEF. Kernforschungszentrum Karlsruhe, KfK-PEF 2, pp 117–131

Kaupenjohann M (1989) Effects of acid rain on soil chemistry and nutrient availability in the soil. In: Schulze ED, Lange OL, Oren R (eds) Forest decline and air pollution. A study of spruce (*Picea abies*) on acid soils. Ecological Studies 77. Springer, Berlin Heidelberg New York, pp 297–340

Kaupenjohann M, Zech W, Hantschel R, Horn R, Schneider BU (1989) Mineral nutrition of forest trees: a regional survey. In: Schulze ED, Lange OL, Oren R (eds) Forest decline and air pollution. A study of spruce (*Picea abies*) on acid soils. Ecological Studies 77. Springer, Berlin Heidelberg New York, pp 282–296

Klemm O (1989) Leaching and uptake of ions through above-ground Norway spruce tree parts. In: Schulze ED, Lange OL, Oren R (eds) Forest decline and air pollution. A study of spruce (*Picea abies*) on acid soils. Ecological Studies 77. Springer, Berlin Heidelberg New York, pp 210–237

Knabe W, Cousen G, Heberle U, Pfeiffer C (1986) Standardisierte Auswertung der immissionsökologischen Waldzustandserfassung. Holz-Zentralblatt 112:681–683

Krivan V, Schaldach G, Hausbeck R, Lüttge U, Kreutzer K (1986) Untersuchungen zur Rolle der Makro- und Mikromineralnährstoffe sowie anderer Elemente bei der Erkrankung von Waldbäumen. In: Projekt Europäisches Forschungszentrum für Maßnahmen zur Luftreinhaltung (ed) 2. Statuskolloquium des PEF. Kernforschungszentrum Karlsruhe, KfK-PEF 4, pp 165–185

Landolt W (1982) Der Einfluß einer praxisnahen SO_2-Begasung auf das $^{14}CO_2$- Fixierungsmuster von Buchen (*Fagus sylvatica* L.) Eur J For Pathol 12:331–339

Lange OL, Heber U, Schulze ED, Ziegler H (1989) Atmospheric pollutants and plant metabolism. In: Schulze ED, Lange OL, Oren R (eds) Forest decline and air pollution. A study of spruce (*Picea abies*) on acid soils. Ecological Studies 77. Springer, Berlin Heidelberg New York, pp 238–273

Linser H, Herwig K (1968) Zusammenhänge zwischen Bewindung, Transpiration und Nährstofftransport bei Lein unter besonderer Berücksichtigung einer variierten Wassergabe und Kalidüngung. Kali-Briefe, Fachgeb 2/2:1–12

Lötschert W, Grosch S (1984) Bleiakkumulation in den Nadeln von *Taxus baccata* L. im Immissionsgebiet von Frankfurt/M. Acta Oecol/Oecol Plant 5:39–47

Marschner H, Richter C (1974) Calciumtransport in Wurzeln von Mais- und Bohnenkeimpflanzen. Plant Soil 40:193–210

Matzner E (1988) Der Stoffumsatz zweier Waldökosysteme im Solling. Habilitationsschrift Forstwissenschaftlicher Fachbereich, Universität Göttingen. Ber des Forschungszentrums Waldökosysteme/Waldsterben, Univ Göttingen, Bd 40, 217 pp

Michaelis W, Schönburg M, Stößel RP (1988) Trocken- und Naßdeposition von Schwermetallen und Gasen. In: Bauch J, Michaelis W (eds) Das Forschungsprogramm Waldschäden am Standort "Postturm", Forstamt Farchau/Ratzeburg. GKSS Forschungszentrum Geesthacht, GKSS 88/E/55, pp 19–59

Michaelis W, Schönburg M, Stößel RP (1989) Deposition of atmospheric pollutants into a North German forest ecosystem. In: Georgii HW (ed) Mechanisms and effects of pollutant-transfer into forests. Kluwer, Dordrecht, pp 3–12

Michaelis W, Rademacher P, Pepelnik R (1991) Stoffflüsse über den Wasserkreislauf innerhalb eines Ökosystems. In: Arbeitsgemeinschaft der Großforschungseinrichtungen (AGF)(ed) Belastung von Böden und Gewässern. Thenée Druck, Bonn, pp 70–73

Michaelis W, Pepelnik R, Theopold F, Rademacher P (1992a) Deposition atmosphärischer Spurenstoffe und Stoffflüsse im Ökosystem Wald. In: Michaelis W, Bauch J (eds) Luftverunreinigungen und Waldschäden am Standort "Postturm", Forstamt Farchau/Ratzeburg. GKSS Forschungszentrum Geesthacht, GKSS 92/E/100, pp 11–59

Michaelis W, Pepelnik R, Prange A (1992b) Application of TXRF in environmental research: Barret CS, Gilfrich JV, Huang TC, Jenkins R, McCarthy GJ, Predecke PK, Ryon R, Smith DK (eds) Advances in X-ray analysis, vol 35B. Plenum Press, New York, pp 953–958

Michaelis W, Pepelnik R, Rademacher P, Riebesell M (1992c) Transfer of atmospheric pollutants into a forest ecosystem. In: Teller A, Mathy P, Jeffers JNR (eds) Responses of forest ecosystems to environmental changes. Elsevier, London, pp 596–597

Murach D (1984) Die Reaktion der Feinwurzeln von Fichten (*Picea abies* Karst.) auf zunehmende Bodenversauerung. Göttinger Bodenkunde Ber 77, 127 pp

Murach D, Rapp C, Ulrich B (1988) Boden- und Feinwurzelinventur auf der Versuchsfläche am Standort "Postturm", Forstamt Farchau/Ratzeburg. In: Bauch J, Michaelis W (eds) Das Forschungsprogramm Waldschäden am Standort "Postturm", Forstamt Farchau/Ratzeburg. GKSS Forschungszentrum Geesthacht, GKSS 88/E/55, pp 189–214

Oberwinkler F, Kottke I, Ritter T, Feil W (1986) Vergleichende Untersuchungen der Feinwurzelsysteme und der Anatomie von Mykorrhizen nach Trockenstress und Düngemaßnahmen. In: Projekt Europäisches Forschungszentrum für Maßnahmen zur Luftreinhaltung (ed) 2. Statuskolloquium des PEF. Kernforschungszentrum Karlsruhe, KfK-PEF 4, pp 315–330

Oren R, Werk KS, Meyer J, Schulze ED (1989) Potentials and limitations of field studies on forest decline associated with anthropogenic pollution. In: Schulze ED, Lange OL, Oren R (eds) Forest decline and air pollution. A study of spruce (*Picea abies*) on acid soils. Ecological Studies 77. Springer, Berlin Heidelberg New York, pp 23–36

Oren R, Zimmermann R (1989) CO_2 assimilation and the carbon balance of healthy and declining Norway spruce stands. In: Schulze ED, Lange OL, Oren R (eds) Forest decline and air pollution. A study of spruce (*Picea abies*) on acid soils. Ecological Studies 77. Springer, Berlin Heidelberg New York, pp 352–369

Panten A (1990) Analyse des Elementgehalts der Kronentraufe unter Fichten in einem geschädigten Waldgebiet. Diplomarbeit, Fachbereich Angewandte Naturwissenschaften, Fachhochschule Lübeck

Rademacher P (1986) Morphologische und physiologische Eigenschaften von Fichten (*Picea abies* [L.] Karst.), Tannen (*Abies alba* Mill.), Kiefern (*Pinus sylvestris* L.) und Buchen (*Fagus sylvatica* L.) gesunder und erkrankter Waldstandorte. GKSS Forschungszentrum Geesthacht, GKSS 86/E/10, 274 pp

Rademacher P, Bauch J, Puls J (1986) Biological and chemical investigations of the wood from pollution-affected spruce (*Picea abies* [L.] Karst.). Holzforschung 40:331–338

Rademacher P, Bauch J, Michaelis W (1988) Einfluß der Elementkonzentration der Bodenlösung auf den Elementgehalt in gesunden und geschädigten Fichten des Standortes "Postturm". In: Bauch J, Michaelis W (eds) Das Forschungsprogramm Waldschäden am Standort "Postturm", Forstamt Farchau/Ratzeburg. GKSS Forschungszentrum Geesthacht, GKSS 88/E/55, pp 215–254

Rademacher P, Ulrich B, Michaelis W (1992) Bilanzierung der Elementvorräte und Elementflüsse innerhalb der Ökosystemkompartimente Krone, Stamm, Wurzel und Boden eines belasteten Fichtenbestandes am Standort "Postturm". In: Michaelis W, Bauch J (eds) Luftverunreinigungen und Waldschäden am Standort "Postturm", Forstamt Farchau/Ratzeburg. GKSS Forschungszentrum Geesthacht, GKSS 92/E/100, pp 149–186

Rehfuess KE (1983) Walderkrankungen und Immissionen – eine Zwischenbilanz: Allg Forstz 38:601–610

Riederer M (1989) The cuticles of conifers: stucture, composition and transport properties. In: Schulze ED, Lange OL, Oren R (eds) Forest decline and air pollution. A study of spruce (*Picea abies*) on acid soils. Ecological Studies 77. Springer, Berlin Heidelberg New York, pp 157–192

Rost-Siebert K (1983) Aluminium-Toxizität und Toleranz an Keimpflanzen von Fichte (*Picea abies* Karst.) und Buche (*Fagus sylvatica* L.) Allg Forstz 38:686–689

Schneider BU, Meyer J, Schulze ED, Zech W (1989) Root and mycorrhizal development in healthy and declining Norway spruce stands. In: Schulze ED, Lange OL, Oren R (eds) Forest decline and air pollution. A study of spruce (*Picea abies*) on acid soils. Ecological Studies 77. Springer, Berlin Heidelberg New York, pp 370–391

Schulte-Bisping H, Murach D (1984) Inventur der Biomasse und ausgewählter chemischer Elemente in zwei unterschiedlich stark versauerten Fichtenjungbeständen im Hils. In: Ber des Forschungszentrums Waldökosysteme/Waldsterben, Univ Göttingen, Bd 2, pp 207–265

Schultz J (1991) Untersuchung von Traufenwasser und Bodenlösung auf Nähr- und Schadelemente - Versuch einer Bilanz. Diplomarbeit, Fachbereich Umwelttechnik, Technische Universität Berlin

Schulz A, Behnke HD (1986) Fluoreszenz- und elektronenmikroskopische Beobachtungen am Phloem von Buchen, Fichten und Tannen unterschiedlichen Schädigungsgrades. In: Projekt Europäisches Forschungszentrum für Maßnahmen zur Luftreinhaltung (ed) 2. Statuskolloquium des PEF. Kernforschungszentrum Karlsruhe, KfK-PEF 4, pp 79-95

Schulze ED, Oren R, Lange OL (1989) Nutrient relations of trees in healthy and declining Norway spruce stands. In: Schulze ED, Lange OL, Oren R (eds) Forest decline and air pollution. A study of spruce (*Picea abies*) on acid soils. Ecological Studies 77. Springer, Berlin Heidelberg New York, pp 392-417

Schütt P, Blaschke H, Hoque E, Koch W, Lang KJ, Schuck HJ (1983) Erste Ergebnisse einer botanischen Inventur des "Fichtensterbens". Forstwiss Centralbl 102:158-166

Stienen H (1985) Struktur und Funktion von Feinwurzeln gesunder und erkrankter Fichten (*Picea abies* [L.] Karst.) unter Wald- und Kulturbedingungen. Dissertation, Fachbereich Biologie, Univ Hamburg, 165 pp

Stienen H, Bauch J (1988) Element content in tissues of spruce seedling from hydroponic cultures simulating acidification and deacidification. Plant Soil 106:231-238

Suske J, Acker G (1989) Endophytic needle fungi: culture, ultrastructural and immunocytochemical studies. In: Schulze ED, Lange OL, Oren R (eds) Forest decline and air pollution. A study of spruce (*Picea abies*) on acid soils. Ecological Studies 77. Springer, Berlin Heidelberg New York, pp 121-136

Tischner R, Kaiser U, Hüttermann A (1983) Untersuchungen zum Einfluß von Aluminium-Ionen auf das Wachstum von Fichtenkeimlingen in Abhängigkeit vom pH-Wert. Forstwiss Centralbl 102:329-336

Uhlmann W, Altner H, Schulze ED, Lange OL (1989) The problem of forest decline and the Bavarian forest toxicology research group. In: Schulze ED, Lange OL, Oren R (eds) Forest decline and air pollution. A study of spruce (*Picea abies*) on acid soils. Ecological Studies 77. Springer, Berlin Heidelberg New York, pp 1-7

Ulrich B (1981) Ökologische Gruppierung von Böden nach ihrem chemischen Bodenzustand. Z Pflanzenernähr Bodenk 144:289-305

Ulrich B (1983) Interaction of forest canopies with atmospheric constituents: SO_2, alkali and earth alkali cations and chloride. In: Ulrich B, Pankrath J (eds) Effects of accumulation of air pollutants in forest ecosystems. D Reidel, Dordrecht, pp 33-45

Ulrich B (1985) Bodenchemische Charakterisierung des wurzelnahen Bodens von Fichten am Standort "Postturm". Bericht über das interdisziplinäre Forschungsprogramm Waldschäden am Standort "Postturm" im Forstamt Ratzeburg, pp 40-42

Ulrich B, Pirouzpanah D, Murach D (1984) Beziehungen zwischen Bodenversauerung und Wurzelentwicklung von Fichten mit unterschiedlich starken Schadsymptomen. Forstarchiv 55:127-134

Wild A, Bode J (1986) Physiologische, biochemische und anatomische Untersuchungen von immissionsbelasteten Fichten verschiedener Standorte. In: Kernforschungsanlage Jülich (KFA)(ed) Wirkung von Luftverunreinigungen auf Waldbäume, Status Sem Jülich 2.-4.12.1985. Spez Ber 369

Zech W, Popp E (1983) Mg-Mangel, einer der Gründe für das Fichten- und Tannensterben in NO-Bayern. Forstwiss Centralbl 102:50-55

12 Concentrations and Deposition of Gaseous Pollutants: SO_2, NO_2, NO, NH_3 and O_3

12.1 Gas Concentrations and Weather Conditions

Depending on the position of the measuring site relative to conurbations, power stations, centres of traffic, industry and agriculture, the atmospheric gas concentrations are subject to the influence of the current weather conditions. Such phenomena which have already been described for the particulate concentrations and the dry deposition of trace elements (Chap. 9) find an analogue above all in the case of the sulphur dioxide immission, which at the "Postturm" investigation site is determined to a high degree by emissions in the brown coal districts in southeast Germany, in spite of the rather great distance involved. In this regard the region Halle-Leipzig-Bitterfeld is of particular importance. The influence should become evident at wind directions from southeast to south-southeast. This is impressively substantiated by Fig. 12.1a which shows the sum of the concentration values per 5° sector element, normalized to the frequency distribution of the wind direction (Michaelis et al. 1990, 1992). The diagram represents the conditions at the early stage of the project, i.e. in the period 1987/1988. Developments after the German reunification will be discussed in Section 12.2.

Since rather frequently circulating winds occur, the impact on the forest ecosystem concentrated primarily upon short-term episodes with partly extreme peak values in the concentration of SO_2. Values up to 720 $\mu g/m^3$ were measured during such events. On the other hand, these episodes offered the chance to study the interrelations between gaseous pollutants and the physiologically important carbon dioxide much more thoroughly than is normally possible under quasi-static conditions (cf. Chap. 13). A typical variation of the SO_2 concentration with changing wind direction is shown in Fig. 12.2 (Michaelis et al. 1988, 1990, 1991, 1992). This plot also reveals the concentration differences at the various heights of the gas inlets.

In the case of the nitrogen oxides NO_x the influence of dense motor traffic in the conurbation of Hamburg is obvious, even though the dependence of the concentration on the wind direction is clearly less pronounced than for SO_2 (Fig. 12.1b). According to the different locations of the sources, the pollution constellation of SO_2 and NO_x changes with the weather conditions in the sense that these gases can reverse their rôles. This is shown in

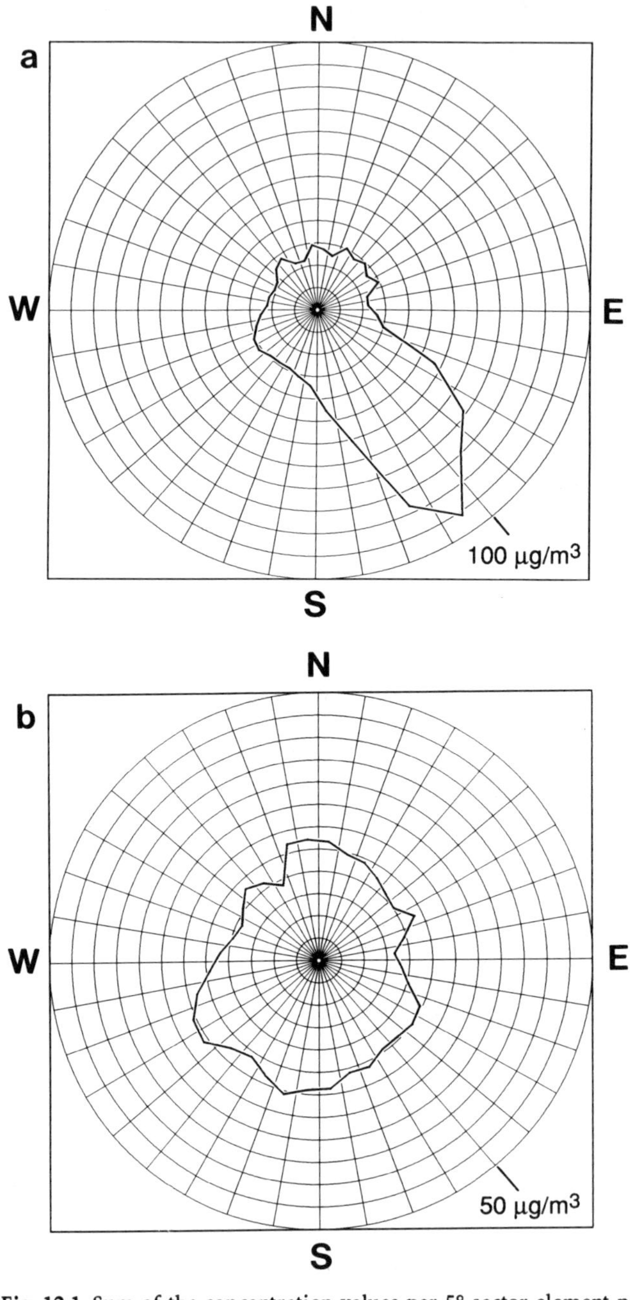

Fig. 12.1. Sum of the concentration values per 5° sector element normalized to the frequency distribution of the wind direction. **a** Sulphur dioxide; **b** nitrogen dioxide. Measuring period: 1987/1988; basis: half-hourly measured values

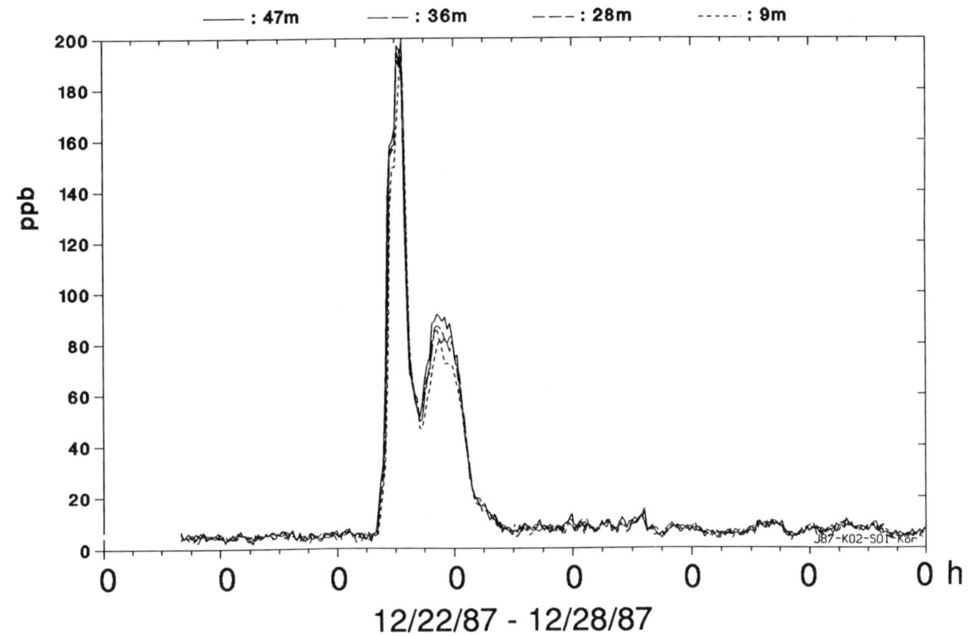

Fig. 12.2. Variation in the sulphur dioxide concentration in ppb during the 52nd calendar week 1987. Half-hourly measured values. For details, see text

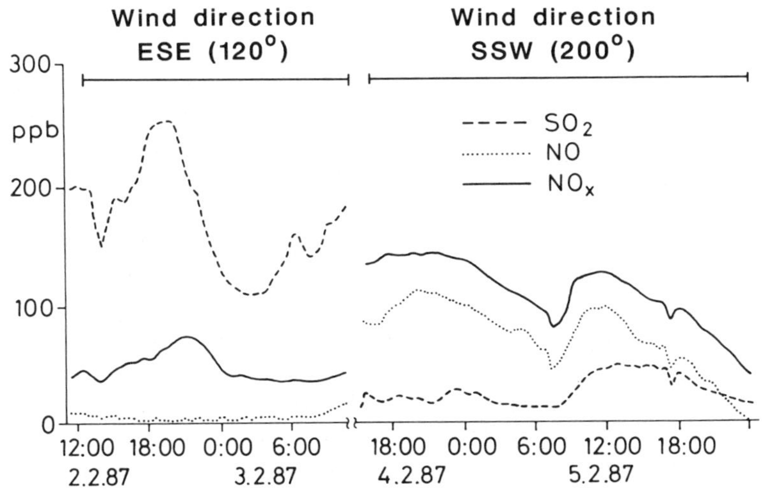

Fig. 12.3. Gas concentrations under changing weather conditions. February 2nd to 5th, 1987

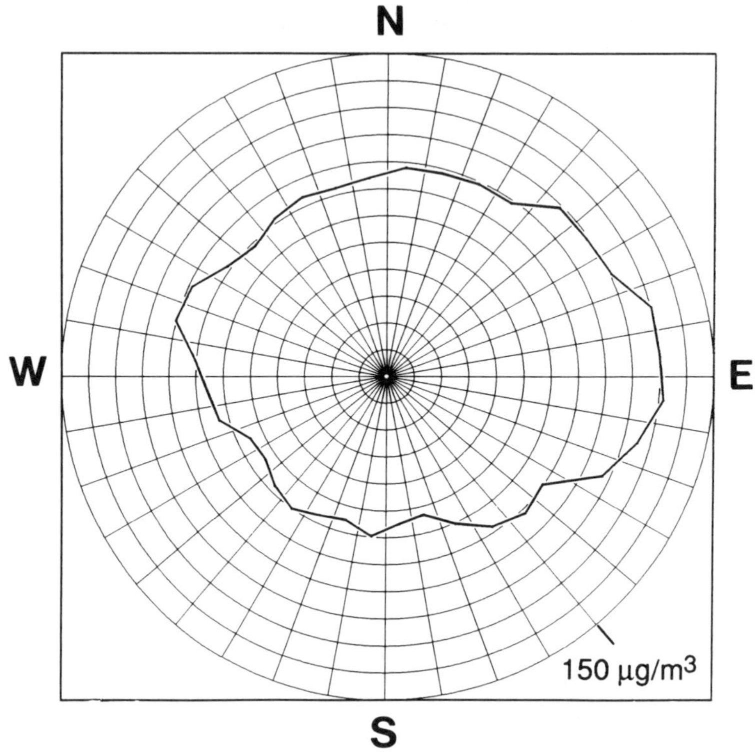

Fig. 12.4. Sum of the O_3 concentration values per 5° sector element normalized to the frequency distribution of the wind direction. Measuring period 1988

Fig. 12.3 where the concentrations of the gases SO_2, NO and NO_x over a period of about 3 days in 1987 are plotted (Michaelis et al. 1989a,b). At the beginning the wind came from a east-southeast to southeast direction. Due to the low traffic density in the former German Democratic Republic the concentrations of the nitrogen oxides were only moderate, whereas the impact of SO_2 turned out to be rather high. When, 2 days later, the wind had shifted to south-southwest, the SO_2 concentration was distinctly reduced as a result of the intensive efforts to restrict the emissions in the Federal Republic. On the other hand, however, the pollution by nitrogen oxides was clearly enhanced due to the dense motor traffic.

The distribution of ozone exhibits a broad maximum at winds originating from easterly directions (Fig. 12.4). The rather low values from the direction of Hamburg may be explained by the partial consumption of O_3 for the oxidation of emitted NO to NO_2. This process is responsible for the fact that in summer the O_3 concentrations in rural regions are often higher than in urban areas.

12.2 Mean Values, Temporal Variations and Long-Term Trends of the Gas Concentrations

The annual mean values determined during the investigation period are summarized in Table 12.1. The data refer to a temperature of 0 °C and a pressure of 1013 hPa.

For SO_2 Prinz and Brandt (1980) as well as Prinz (1982) recommended a maximum annual mean of 50 μg/m³, in order to ensure utmost protection of the forests. Table 12.1 reveals that this value was not exceeded in the course of the whole measuring period. However, the question arises whether short-term peak concentrations can cause damage to the trees. This problem will be taken up again in Chapter 13. In the case of nitrogen oxides, the World Health Organization (WHO 1985) recommended a maximum annual mean value of 30 μg/m³ referring explicitly to the synergism with SO_2, since nitrogen compounds alone are considered to be less harmfull than SO_2. Again, however, the influence of temporarily extreme concentrations should be borne in mind (cf. Chap. 13).

The situation is quite different in the case of O_3. Here, the annual mean values were above the recommended limits over the entire measuring period. Long-term exposure experiments caused first damage to young needles at concentrations above 60 μg/m³. The guide value of the World Health Organization (WHO 1985) amounts to 60 μg/m³ as the mean during the vegetation period. A distinct reduction of photosynthesis was established in the case of deciduous trees at concentrations of 100 μg/m³ under a several days' to several weeks' impact (Forschungsbeirat Waldschäden/Luftverunreinigungen 1986). As will be shown below, the median values at the "Postturm" site during the months April to July were found to be above 120 μg/m³ with very few exceptions. Ozone therefore requires particular attention.

The data summarized in Table 12.1 in part clearly exceed findings from South German forest decline areas. For the "Kälbelescheuer" in the southern Black Forest, for instance, the following values are reported in the literature (Kilz 1987): SO_2 10 μg/m³; NO_2 9 μg/m³ and O_3 63 μg/m³. These data are valid for the year 1985.

Table 12.1. Annual mean values of the concentrations of SO_2, NO_2, NO, O_3 and NH_3 in μg/m³

	1987[a]	1988	1989	1990	1991	1992[b]
SO_2	28.5	31.2	24.7	17.5	20.8	13.5
NO_2	24.0	21.0	23.3	24.0	22.1	19.9
NO	2.9	3.0	5.2	3.6	4.6	4.6
O_3	72.1	85.4	74.5	67.5	68.9	76.6
NH_3	-	4.6	2.9	3.7	4.9	2.8

[a] February to December only
[b] January to July only

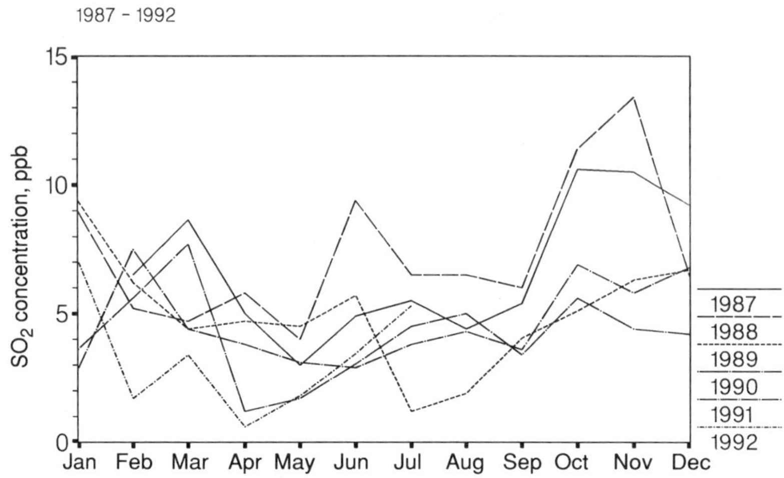

Fig. 12.5. Seasonal variation of the SO_2 concentration in ppb plotted as the monthly median value over the measuring period 1987 to 1992

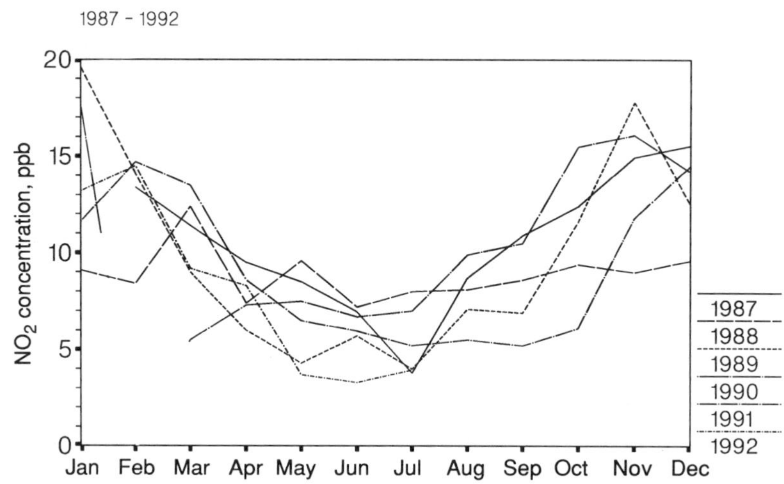

Fig. 12.6. Seasonal variation of the NO_2 concentration in ppb plotted as the monthly median value over the measuring period 1987 to 1992

Annual mean values only insufficiently reflect the relevant factors which influence the degree of pollution. First indications are delivered by the seasonal and diurnal variations in the concentrations. These courses are shown in Figs. 12.5 to 12.7 and 12.9 to 12.11, respectively, for the gases SO_2, NO_2 and O_3. The plots represent the monthly and hourly median values, respectively. In order to illustrate the influence of photochemistry on the interrelations between nitrogen oxides, ozone and hydrocarbons or other carbon

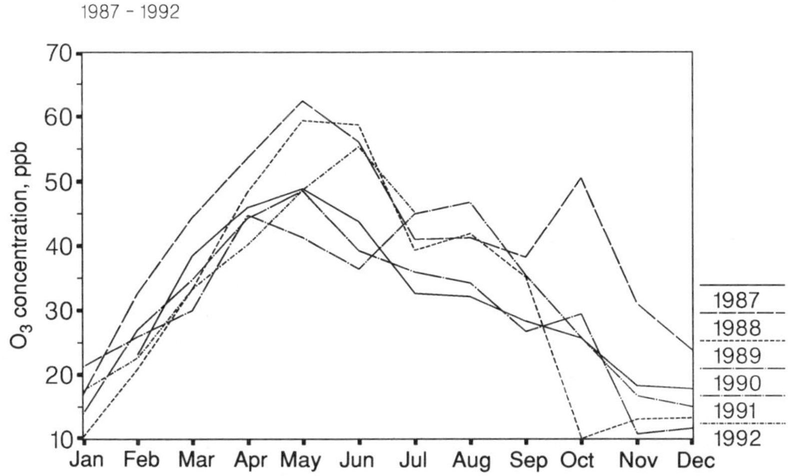

Fig. 12.7. Seasonal variation in the O_3 concentration in ppb plotted as the monthly median value over the measuring period 1987 to 1992

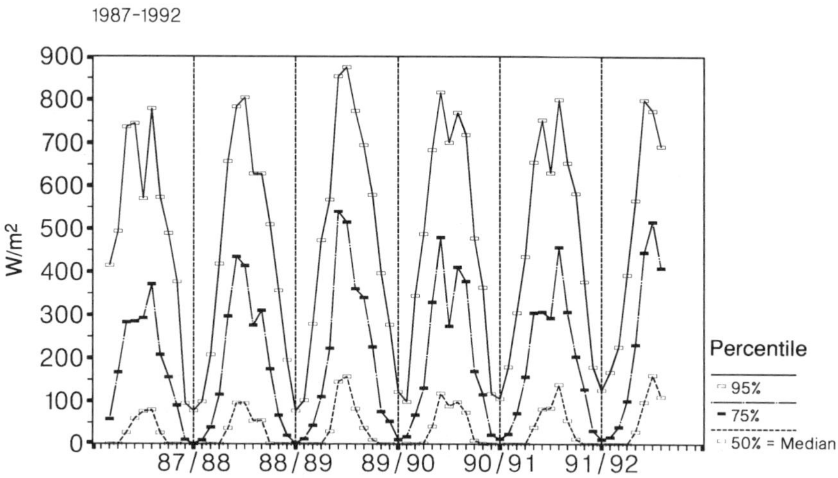

Fig. 12.8. Monthly percentiles of the global radiation in W/m^2 during the measuring period 1987 to 1992

compounds (Kley and Volz-Thomas 1990), Figs. 12.8 and 12.12 present the seasonal and the diurnal variation of the global radiation. In the first case, the plot of monthly percentiles was chosen, since the median values in the period October to March approach zero due to the day-night rhythm. The diurnal course of the global radiation is presented in the form of the hourly median values of 2 months in each case.

On the average, the concentration of SO_2 shows clearly enhanced values during the period October to March. Also in the case of NO_2, a pronounced

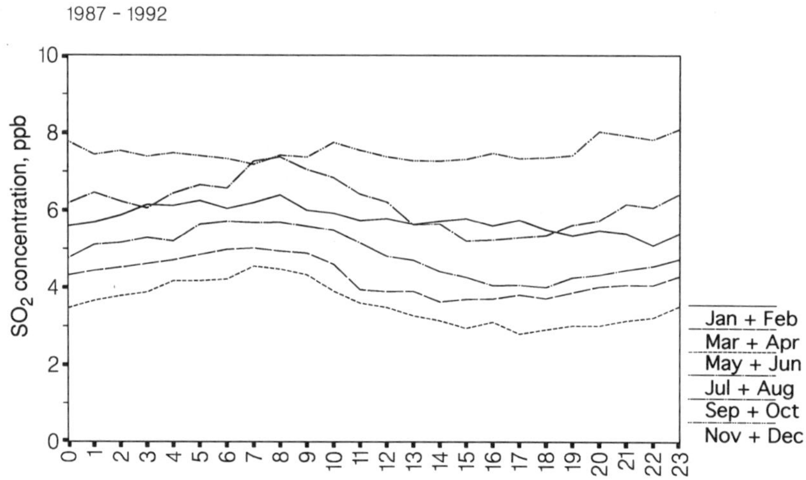

Fig. 12.9. Diurnal variation in the SO_2 concentration in ppb plotted as the hourly median value over 2 months each. Measuring period 1987 to 1992

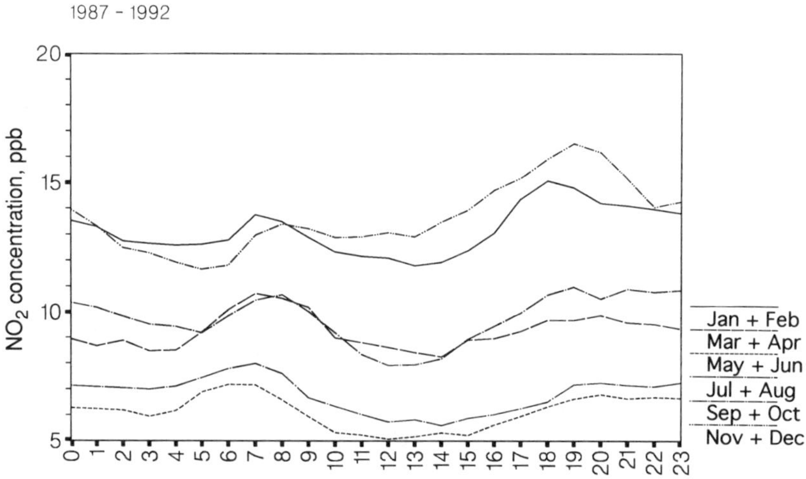

Fig. 12.10. Diurnal variation in the NO_2 concentration in ppb plotted as the hourly median value over 2 months each. Measuring period 1987 to 1992

seasonal variation is evident with maxima in the winter half-year and minima in the summer. The NO concentration exhibits strong fluctuations with a frequent occurrence of peaks in the period October to April (Fig. 12.13). As expected, O_3 shows a distinct parallelism to the course of the global radiation (Figs. 12.7 and 12.8).

The diurnal variation of the SO_2 concentration is only weak (Fig. 12.9). Nitrogen dioxide is characterized by two remarkable maxima, one in the early morning, the other during the evening hours (Fig. 12.10). A compari-

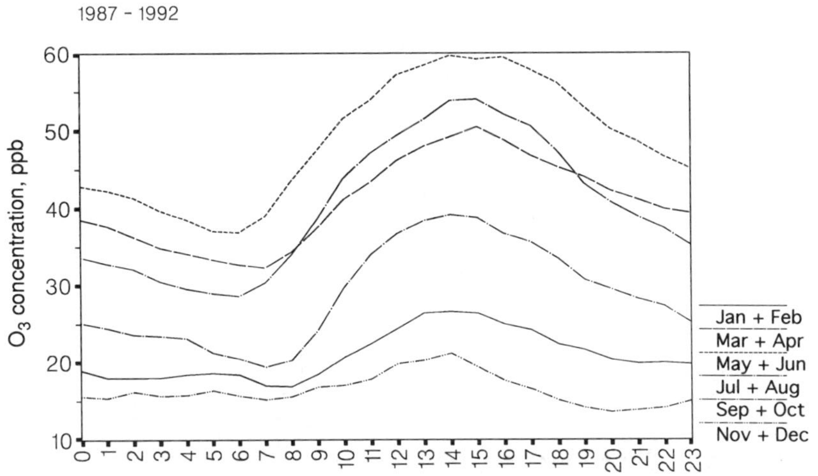

Fig. 12.11. Diurnal variation of the O_3 concentration in ppb plotted as the hourly median value over 2 months each. Measuring period 1987 to 1992

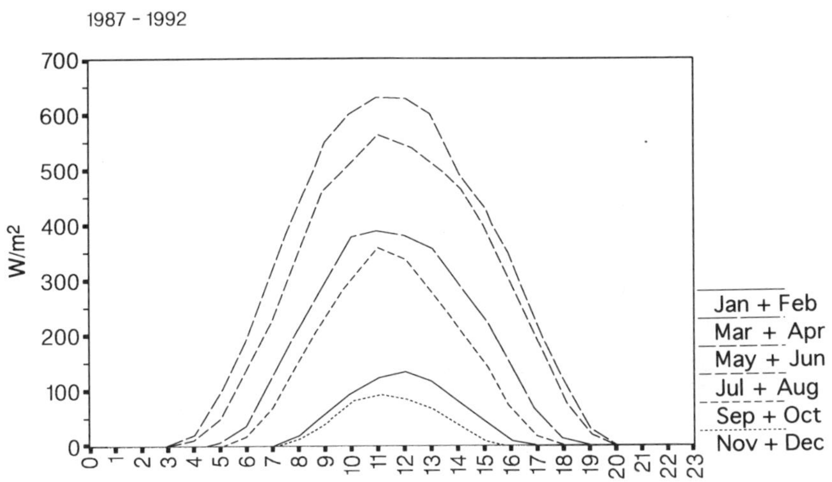

Fig. 12.12. Diurnal variation in the global radiation in W/m^2. Plotted as the hourly median value over 2 months each. Measuring period 1987 to 1992

son of the diurnal variation of the O_3 concentration with that of the global radiation again indicates the influence of the photochemistry (Figs. 12.11 and 12.12). The maxima of O_3 are shifted by a few hours, since in rural regions the formation of this oxidant occurs more slowly than in urban areas due to the higher percentage of only slowly oxidizable hydrocarbons. However, at weather conditions with wind from southwest the course of the concentration might also be influenced by the conurbation of Hamburg where above all the rapidly oxidizable hydrocarbons react.

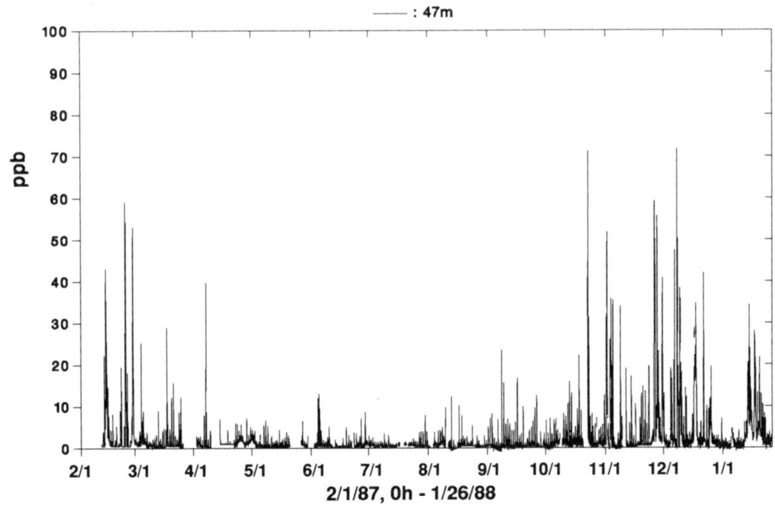

Fig. 12.13. Seasonal variation in the NO concentration in ppb. Time resolution 30 min. Measuring period February 1987 to January 1988

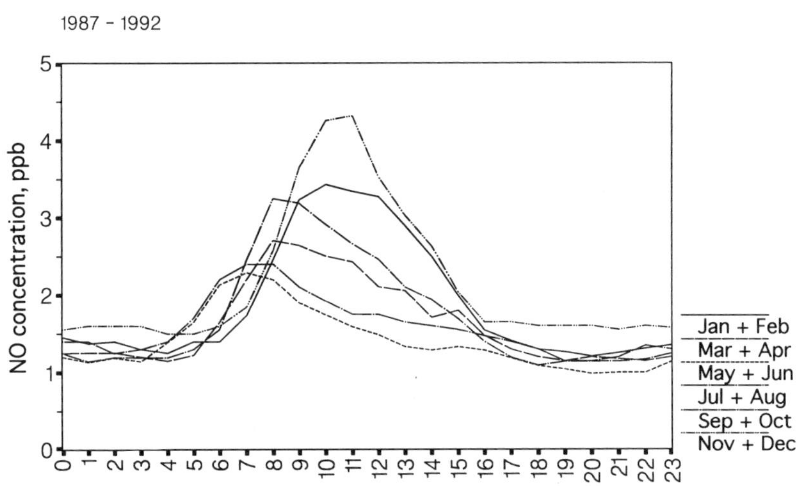

Fig. 12.14. Diurnal variation in the NO concentration in ppb plotted as the hourly median value over 2 months each. Measuring period 1987 to 1992

Of particular interest are also the diurnal variations of NO and CO_2 which are plotted in Figs. 12.14 and 12.15. Nitrogen monoxide shows a pronounced maximum in the course of the forenoon, whereby height and instant depend on the time of the year. In the early morning at low radiation intensity, the emission of nitrogen oxides and hydrocarbons increases. With rising radiation NO_2 undergoes photolysis and the produced oxygen atoms

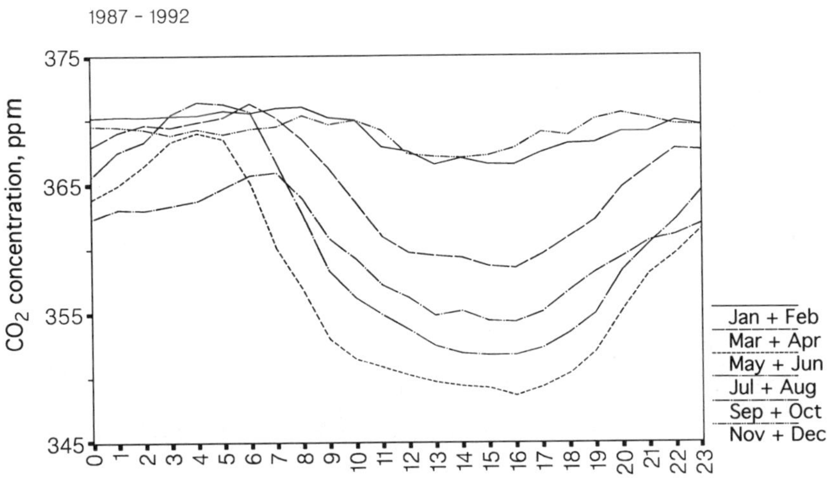

Fig. 12.15. Diurnal variation in the CO_2 concentration in ppm plotted as the hourly median value over 2 months each. Measuring period 1987 to 1992

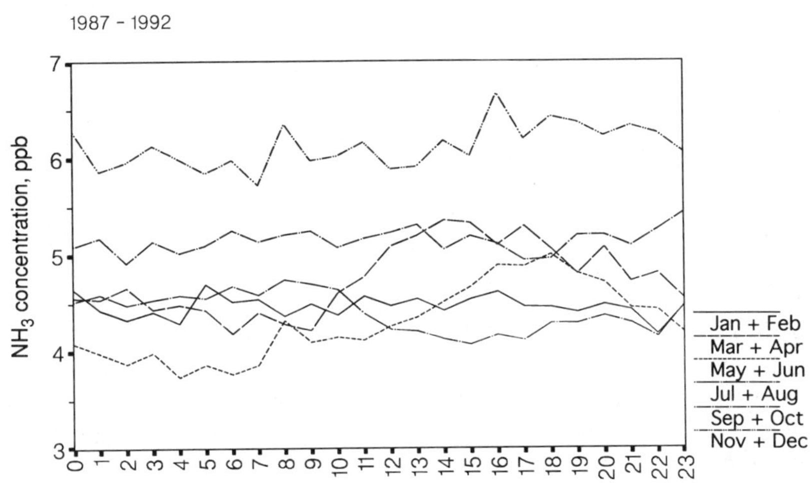

Fig. 12.16. Diurnal variation in the NH_3 concentration in ppb plotted as the hourly median value over 2 months each. Measuring period 1987 to 1992

react with molecular oxygen to form ozone. With increasing O_3 concentration NO is degraded by oxidation thus re-forming NO_2. In the presence of hydrocarbons or other carbon compounds, e.g. aldehydes and ketones, it comes to a net production of O_3. The main sources of the precursors are incomplete combustion of fossile fuels, evaporative losses and industrial emissions, but also biogenic sources have to be considered. An example is isoprene (C_5H_8) which is released by certain deciduous trees.

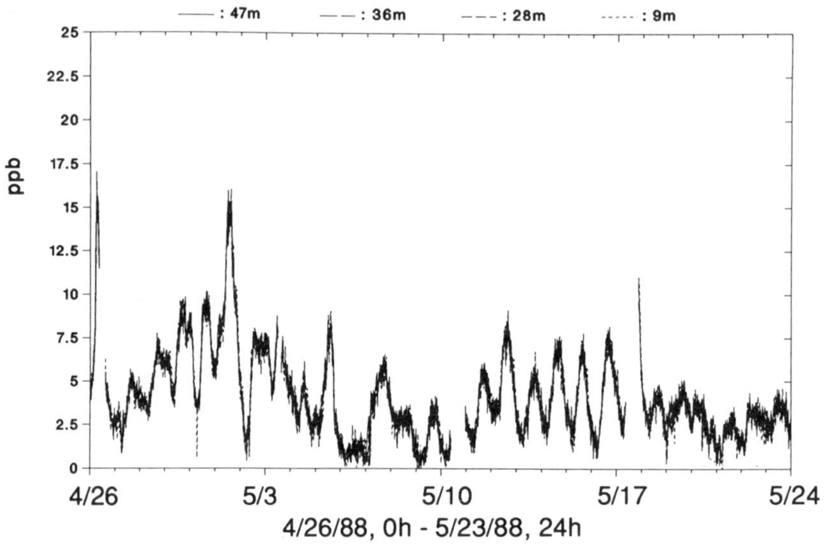

Fig. 12.17. Temporal variation in the NH_3 concentration in the measuring period 26th April to 23rd May 1988. Time resolution 30 min

Carbon dioxide exhibits the well-known day-night rhythm (Fig. 12.15), clearly developed during the vegetation period and only weakly from November to February. The diurnal variation of the NH_3 concentration is practically zero in the course of two thirds of the year. Only during the months March to June is there a distinct variation with maxima in the afternoon and the early evening (Figs. 12.16 and 12.17).

From the gases listed in Table 12.1 only SO_2 shows a clear long-term trend with a decrease of about 50% during the investigation period. This trend did not significantly start before 1990 which is in good agreement with the results obtained for the dry deposition of sulphur via airborne particulates (Chap. 9). The reduction is mainly due to the diminution of the SO_2 emissions in eastern Germany. This statement is substantiated in Fig. 12.18 which shows the sum of the concentration values per 30° sector element normalized to the frequency distribution of the wind direction during the period 1987 to 1992. The measurements of the other gases do not reveal any significant decrease in the concentrations. In the case of ozone, a rise in the tropospheric concentration which has often been reported in the literature is not apparent from Table 12.1. Most probably, the evident increase in the former decades has come to a standstill as a consequence of the stabilization of the nitrogen oxide and hydrocarbon emissions. The fundamentals of atmospheric chemistry with respect to photochemical oxidants are comprehensively treated in the literature by, among others, Seinfeld (1980), Becker et al. (1985), Kley and Volz-Thomas (1990), Kley et al. (1990, 1993), Mihelcic et al. (1993) and Volz-Thomas et al. (1993).

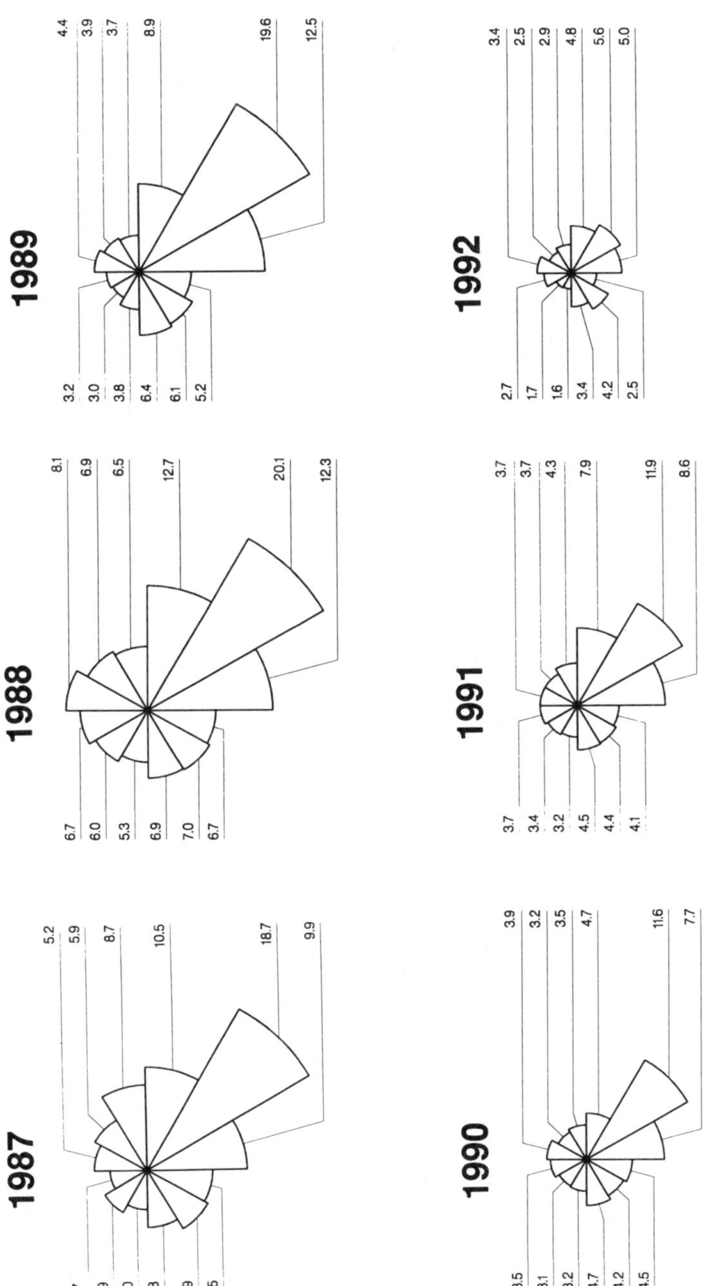

Fig. 12.18. Sum of the SO_2 concentration values per 30° sector element normalized to the frequency distribution of the wind direction. Long-term trend in the measuring period 1987 to 1992. The figures indicate the mean concentration in ppb

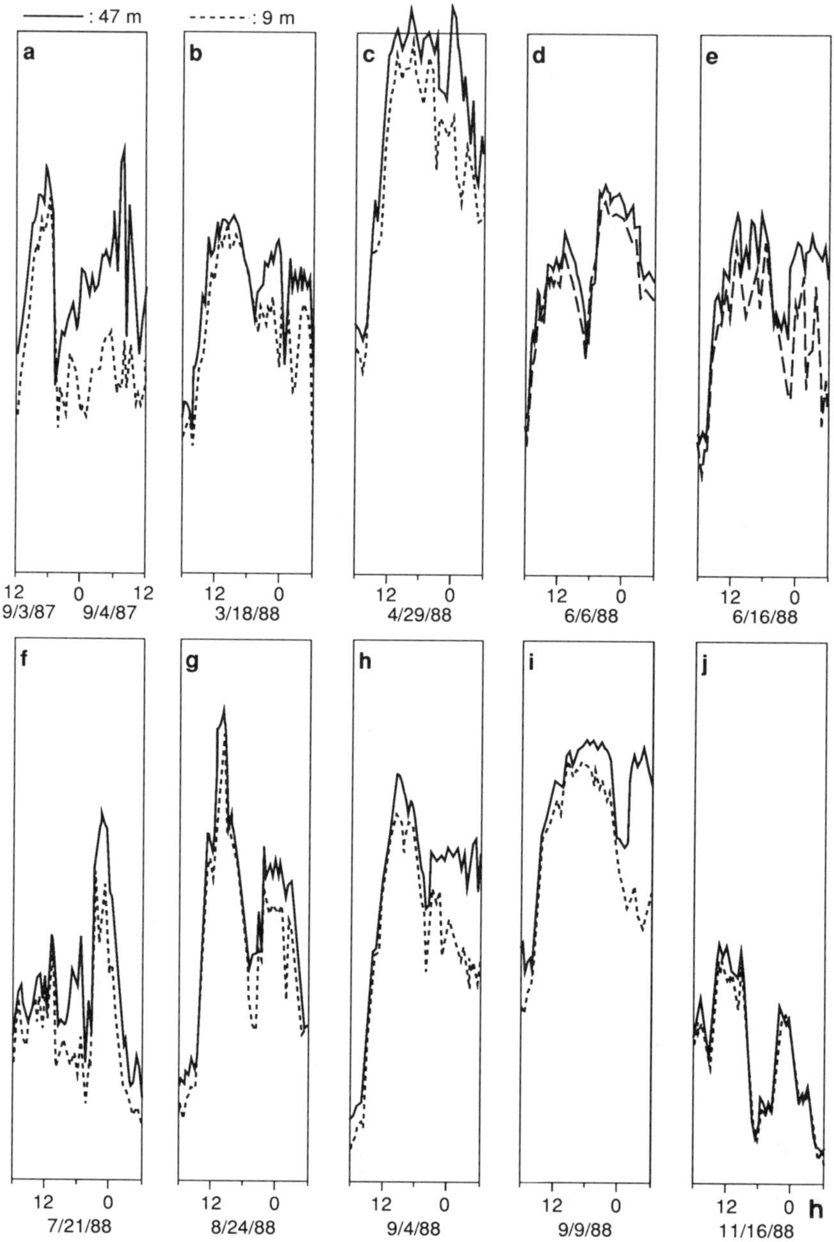

Fig. 12.19.a-j Examples of nocturnal ozone maxima. Concentrations at 47 m height (*full curve*) and at 9 m (*dashed curve*), full scale 100 ppb

12.3 Nocturnal Ozone Maxima

The data presented in the preceding section with respect to the diurnal variation of the O_3 concentration essentially reflect the prevalent conditions with maxima in the afternoon which lag behind the peak of global radiation by a few hours, and low nighttime values due to the absence of solar radiation and trapping of ozone reducers near the surface in a rather shallow inversion layer. The occurrence of anomalies with maxima also during the night is hidden when only mean values are considered, though such episodes are not very scarce. Some examples of nocturnal O_3 maxima observed at the "Postturm" site are compiled in Fig. 12.19 where the courses of the concentrations measured at two heights, 9 and 47 m, are plotted over a period of 24 hours in each case. Mostly these events occur around midnight and the maximum values are usually lower than those in the daytime. However, there are also episodes with higher values during the night (Fig. 12.19a,d,f). In other cases, particularly in summer, both maxima merge into each other (Fig. 12.19c). During this period, it may happen that several times per week nocturnal rises in the O_3 concentration occur. Numerous events are also observed in winter though less pronounced than those occurring during the summer. Sometimes there are two or even three maxima per night (Fig. 12.19j). Generally, the nocturnal ozone maxima are characterized by marked gradients in the concentration (Fig. 12.19).

Most probably these anomalies have to be ascribed to meteorological processes (Samson 1978; Winkler 1980). Above the surface inversion, O_3 concentrations are assumed to remain relatively high throughout the night. The rise in the concentration near the surface may be caused by the development of a nocturnal low-level wind jet just above the surface inversion layer. This jet is probably, at least in part, the result of the decoupling of the boundary layer from the effects of surface friction. A theory of this phenomenon has been propounded by Blackadar (1957). The jet produces turbulence in the inversion layer and conveys ozone-rich air towards the surface.

12.4 Dry Deposition: Mean Values, Temporal Variations and Long-Term Trends

As described in Chapter 6, the turbulent fluxes in the atmosphere-forest boundary layer were determined using the gradient method. The experimental conditions and the demands on the measuring technique are exemplified in Fig. 12.20 by means of the differences between the concentration values of ozone measured at the various heights. For the other gases, the conditions are often even more critical.

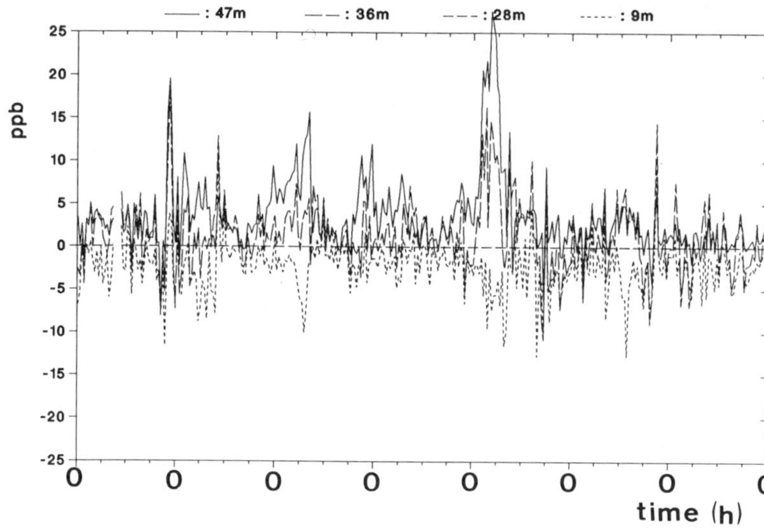

Fig. 12.20. Difference in the O_3 concentration values measured at 47, 36 and 9 m to that at 28 m. Data from the 36th calendar week 1988. Time resolution 30 min

Table 12.2. Annual mean values of the turbulent fluxes in the boundary layer atmosphere – forest ecosystem. Data for SO_2, NO_2, NO and O_3 in µg/m²s, CO_2 data in mg/m²s

	1987[a)]	1988	1989	1990	1991	1992[b)]
SO_2	– 0.15	– 0.15	– 0.14	– 0.069	– 0.046	(– 0.024)
NO_2	0.003	0.011	0.011	0.016	0.017	0.005
NO	0.003	– 0.001	– 0.002	– 0.004	– 0.001	0.0001
O_3	– 0.48	– 0.47	– 0.42	– 0.38	– 0.34	– 0.028
CO_2	0.032	0.062	– 0.002	0.026	0.022	0.017

a) February to December only
b) January to July only

The annual mean values of the fluxes obtained for SO_2, NO_2, NO, O_3 and CO_2 during the investigation period are summarized in Table 12.2 (Michaelis et al . 1992; Michaelis and Theopold 1993). Throughout, negative values for the dry deposition occur in the case of SO_2 and O_3. A negative sign here indicates fluxes into the ecosystem, whereas a positive sign means a release into the atmosphere. By analogy with Table 12.1, sulphur dioxide reveals a pronounced decreasing tendency. As regards the result for 1992, however, the restriction must be made that this mean does not include measurements for the months August to December and thus for the beginning of the heating period (cf. seasonal variation in Fig. 12.5). When assessing the data in Table 12.2, it must also be taken into account that, compared to Table 12.1, in addition to the errors of the gas measurement, uncertainties in the determination of the turbulent exchange coefficient enter into the re-

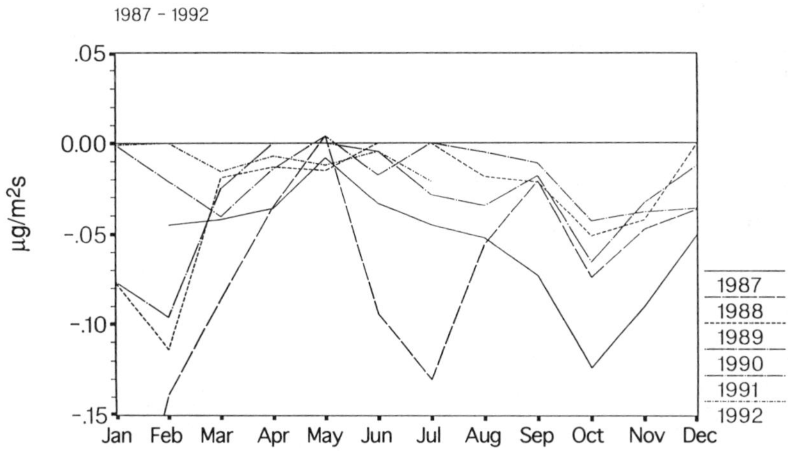

Fig. 12.21. Seasonal variation in the SO_2 dry deposition in $\mu g/m^2s$ plotted as the monthly median value over the measuring period 1987 to 1992

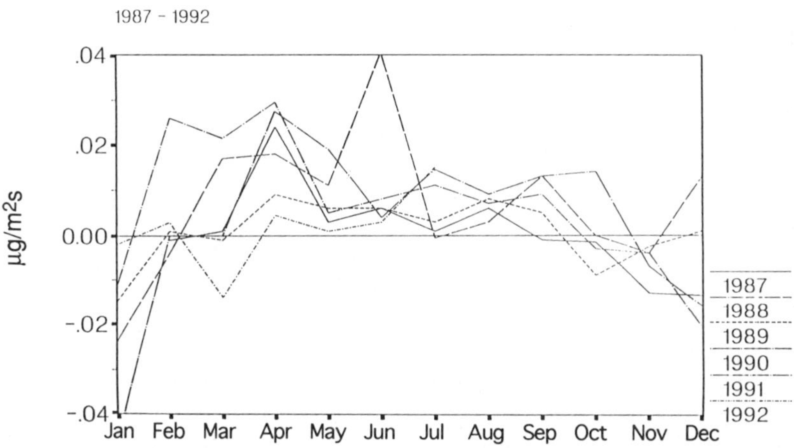

Fig. 12.22. Seasonal variation in the NO_2 flux in $\mu g/m^2s$ plotted as the monthly median value over the measuring period 1987 to 1992

sults. Considering this aspect, the fluxes of NO_2 and NO do not deviate significantly from zero. A similar conclusion can be drawn in the case of CO_2; the net balance is, contrary to the expectations, practically zero as well. On the other hand, there is a pronounced dry deposition of O_3. In accordance with Section 12.2 no increase is evident during the investigation period. Fluxes of NH_3 could not be determined due to the long response time of the analyser (cf. Sect. 5.2). The impact of the gaseous pollutants on the forest ecosystem will be discussed in greater detail in Chapter 13.

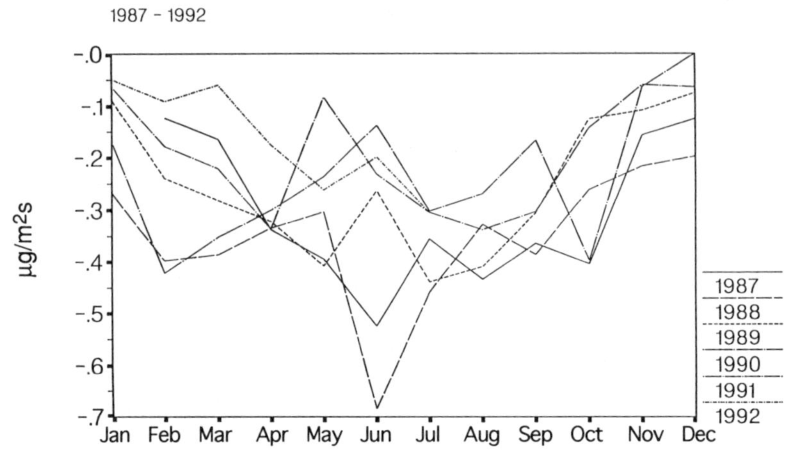

Fig. 12.23. Seasonal variation in the O_3 dry deposition in µg/m²s plotted as the monthly median value over the measuring period 1987 to 1992

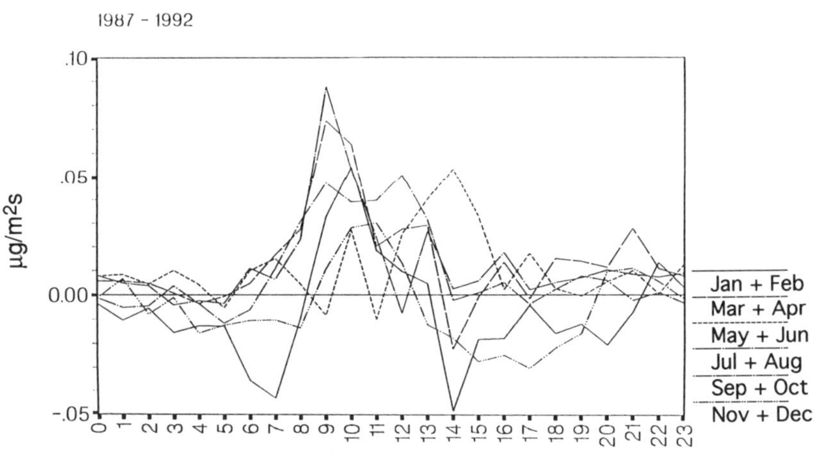

Fig. 12.24. Diurnal variation in the NO_2 flux in µg/m²s plotted as the hourly median value over 2 months each. Measuring period 1987 to 1992

The turbulent fluxes in part show pronounced seasonal and diurnal variations. In Figs. 12.21 to 12.23, the monthly median values for the gases SO_2, NO_2 and O_3 are illustrated. As expected, the deposition of SO_2 is on the average during the summer lower than in the winter period. Nitrogen dioxide exhibits slightly positive values from spring to autumn and in the winter a tendency towards negative fluxes. In the case of O_3 the median value is throughout negative with a distinct maximum of the deposition during the summer period. As shown by the monthly percentiles, the fluxes of all gases

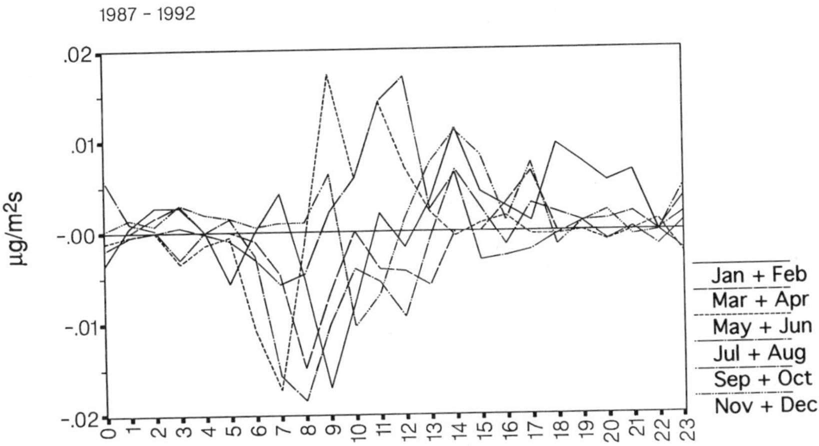

Fig. 12.25. Diurnal variation in the NO flux in µg/m²s plotted as the hourly median value over 2 months each. Measuring period 1987 to 1992

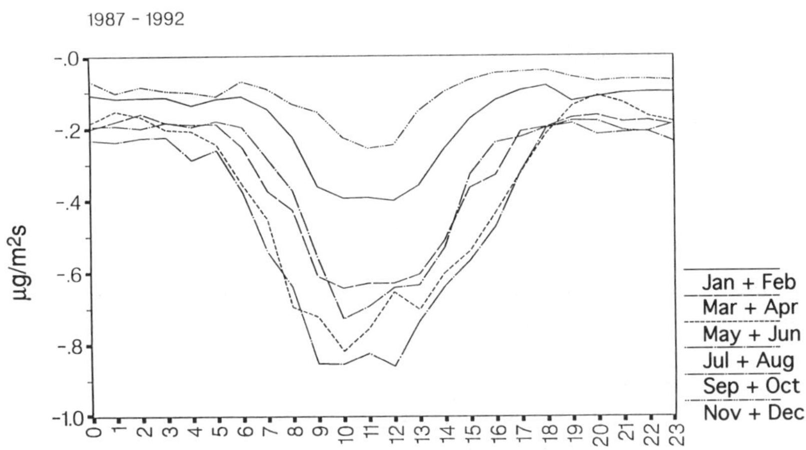

Fig. 12.26. Diurnal variation in the O_3 dry deposition in µg/m²s plotted as the hourly median value over 2 months each. Measuring period 1987 to 1992

are characterized by rather broad distributions. Due to the great fluctuations nitrogen monoxide does not reveal a significant annual course.

In Figs. 12.24 to 12.27, the diurnal variations of the gases NO_2, NO, O_3 and CO_2 are plotted as the hourly median value of 2 months. All these gases show more or less distinct courses. In the case of the nitrogen oxides, the data may suggest opposite signs of the fluxes in the morning. The variations in the O_3 and CO_2 fluxes are at first sight in accordance with the expectations though the overall balance of CO_2 appears to be quite unsatisfactory. Sulphur dioxide does not exhibit a significant dependence on the time of the day which is obvious on the basis of Fig. 12.9.

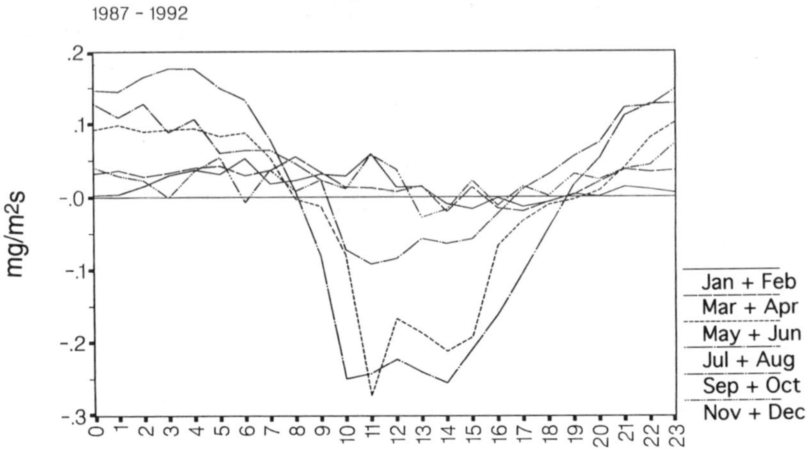

Fig. 12.27. Diurnal variation in the CO_2 flux in mg/m^{2}s plotted as the hourly median value over 2 months each. Measuring period 1987 to 1992

12.5 Deposition Velocities

If the equations for the concentration method [Eq.(1), Sect. 6.1] and the gradient method [Eq.(2), Sect. 6.1] are combined in the form:

$$c_{z_o} v_D = K \frac{dc}{dz},$$

the deposition velocities v_D of gases may be derived by means of the methodology used in the present study. From Tables 12.1 and 12.2 the following annual mean values result: 0.44 cm/s for SO_2 and 0.57 cm/s for O_3.

A direct comparison of these findings with data reported in the literature is highly problematic, because (1) the experimental conditions were often insufficiently defined, (2) different properties of the vegetation and incomparable meteorological effects influenced the measurements, or (3) the data were obtained only by short-term experiments. Therefore the results are spread over more than one order of magnitude. For a mainly wooded area, McMillen (pers. comm.) derived from model considerations seasonal variations of between 0.1 and 0.4 cm/s in the deposition velocity for both SO_2 and O_3. Using eddy correlation and profile flux methods, Droppo (1985) obtained a mean value of 0.6 cm/s for O_3 during five afternoon hours on a summer day. Derwent and Hov (1979, 1980; see also Becker et al. 1985) used the deposition velocities 0.8 cm/s for SO_2 and 0.6 cm/s for O_3 in model calculations. Markedly higher values were quoted by Jonas (1983) who specified 1.54 cm/s for SO_2 in the case of pinewood and mixed forests as well as 7.6 cm/s (single value) for the deposition of O_3 into an oak forest.

References

Becker KH, Fricke W, Löbel J, Schurath (1985) Formation, transport, and control of photochemical oxidants. In: Guderian R (ed) Air pollution by photochemical oxidants. Formation, transport, control, and effects on plants. Ecological Studies 52. Springer, Berlin Heidelberg New York, pp 3–125

Blackadar AK (1957) Boundary layer wind maxima and their significance for the growth of nocturnal inversions. Bull Am Meterol Soc 38:238–290

Derwent RG, Hov Ø (1979) Computer modeling studies of photochemical air pollution formation in north-west Europe. Environ Med Sci Div, AERE Harwell, UK

Derwent RG, Hov Ø (1980) Computer modeling studies of the impact of vehicle exhaust emission controls on photochemical air pollution in the United Kingdom. Environ Sci Technol 14:1360–1366

Droppo JG (1985) Concurrent measurements of ozone dry deposition using eddy correlation and profile flux methods. J Geophys Res 90:2110–2118

Forschungsbeirat Waldschäden/Luftverunreinigungen der Bundesregierung und der Länder (1986). 2. Bericht. Karl Elser Druck GmbH, Mühlacker

Jonas R (1983) Ablagerungsgeschwindigkeit von Aerosolen und Gasen auf Vegetation und ebene Oberflächen. In: Arbeitsgemeinschaft der Großforschungseinrichtungen (AGF)(ed) Luftreinhaltung - Luftverschmutzung. Thenée Druck, Bonn, pp 24–26

Kilz E (1987) Charakterisierung des Standortes Kälbelescheuer. In: Siefermann-Harms D, Kilz E (eds) Interdisziplinärer PEF-Forschungsschwerpunkt Kälbelescheuer/Südschwarzwald. Kernforschungszentrum Karlsruhe, KfK-PEF 10

Kley D, Volz-Thomas A (1990) Die Belastung der Umwelt durch troposphärisches Ozon. In: Jahresbericht 1990. Forschungszentrum Jülich GmbH, Jülich

Kley D, Geiss H, Heil T, Holzapfel C (1990) Ozon in Deutschland. Die Belastung durch Ozon in ländlichen Gebieten im Kontext der neuartigen Waldschäden. Monographien des Forschungszentrums Jülich GmbH, Band 2. D. Gehler, Graphische Kunstanstalt Düren

Kley D, Geiss H, Klemp D, Kramp F, Su Y, Volz-Thomas A (1993) The importance of hydrocarbon measurements. In: Borrell PM (ed) Proceedings of EUROTRAC Symposium '92. SPB Academic Publishing bv, The Hague, pp 70–79

Michaelis W, Theopold F (1993) Deposition atmosphärischen Ozons und ihre Wirkung auf ein Waldökosystem. In: Arbeitsgemeinschaft der Großforschungseinrichtungen (AGF)(ed) Atmosphärisches Ozon. Prozesse und Wirkungen. Thenée Druck, Bonn, pp 25–27

Michaelis W, Schönburg M, Stößel RP (1988) Trocken-und Naßdeposition von Schwermetallen und Gasen. In: Bauch J, Michaelis W (eds) Das Forschungsprogramm Waldschäden am Standort "Postturm", Forstamt Farchau/Ratzeburg. GKSS Forschungszentrum Geesthacht, GKSS 88/E/55, pp 19–59

Michaelis W, Schönburg M, Stößel RP (1989a) Deposition of atmospheric pollutants into a North German forest ecosystem. In: Georgii HW(ed) Mechanisms and effects of pollutant-transfer into forests. Kluwer, Dordrecht, pp 3–12

Michaelis W, Schönburg M, Stößel RP (1989b) Schadstofftransfer in der Grenzschicht Atmosphäre-Vegetation. In: Arbeitsgemeinschaft der Großforschungseinrichtungen (AGF) (ed) Wechselwirkung Atmosphäre-Biosphäre. Thenée Druck, Bonn, pp 29–33

Michaelis W, Pepelnik R, Rademacher P, Riebesell M (1990) Wechselwirkung zwischen Luftschadstoffen und Vegetation. In: GKSS Jahresbericht 1990. GKSS Forschungszentrum Geesthacht, pp 42–55

Michaelis W, Pepelnik R, Rademacher P, Riebesell M (1991) Transfer of atmospheric pollutants into a forest ecosystem. In: Teller A, Mathy P, Jeffers JNR (eds) Responses of forest ecosystems to environmental changes. Elsevier, London, pp 596–597

Michaelis W, Pepelnik R, Theopold F, Rademacher P (1992) Deposition atmosphärischer Spurenstoffe und Stoffflüsse im Ökosystem Wald. In: Michaelis W, Bauch J (eds) Luftverunreinigungen und Waldschäden am Standort "Postturm", Forstamt Farchau/Ratzeburg. GKSS Forschungszentrum Geesthacht, GKSS 92/E/100, pp 11–59

Mihelcic D, Klemp D, Müsgen P, Pätz HW, Volz-Thomas A (1993) Simultaneous measurements of peroxy and nitrate radicals at Schauinsland. J Atmos Chem 16:313–335

Prinz B (1982) Wirkungen von Luftverunreinigungen auf Pflanzen und Möglichkeiten zum verbesserten Schutz der Vegetation in der Bundesrepublik Deutschland. In: Rat von Sachverständigen für Umweltfragen (ed) Materialien zu Energie und Umwelt. Kohlhammer-Verlag, Stuttgart

Prinz B, Brandt CJ (1980) Study on the impact of the principal atmospheric pollutants on the vegetation. In: Commission of the European Community (ed) EUR 6644 EN, Brussels

Samson PJ (1978) Nocturnal ozone maxima. Atmos Environ 12:951–955

Seinfeld JH (1980) Lectures in atmospheric chemistry. Monograph series, vol 76, no 12. American Institute of Chemical Engineers, New York

Volz-Thomas A, Flocke F, Garthe HJ, Geiss H, Gilge S, Heil T, Kley D, Klemp D, Kramp F, Mihelcic D, Pätz HW, Schultz M, Su Y (1993) Photo-oxidants and precursors at Schauinsland, Black Forest. In: Borrell PM (ed) Proceedings of EUROTRAC Symposium '92. SPB Academic Publishing bv, The Hague, pp 98–103

Winkler P (1980) Störung der nächtlichen Grenzschicht. Meteorol Rundsch 33:90–94

World Health Organization (1985) Air quality guidelines – ecological effects of air pollutants. ICP/CEH 902/m 71 (S), Geneva

13 Impact of Gaseous Pollutants on the Forest Ecosystem

13.1 Interrelations Between the Concentrations of Pollutants and Carbon Dioxide

Of particular interest are the findings that high concentrations of sulphur dioxide, nitrogen monoxide, nitrogen dioxide and ozone cause marked alterations in the concentration of carbon dioxide (Michaelis et al. 1988, 1989a,b, 1990, 1991, 1992; Michaelis and Theopold 1993). The temporal behaviour and the dimensions of this effect depend on the pollutant considered and its concentration. Such episodes occur at any time of the day and also during the vegetation rest. A general feature is an increase in the concentration of CO_2. The effect is superimposed upon the normal diurnal and seasonal variation. The amplitude often markedly exceeds the usual rise during the night, and the day-night rhythm can be changed over several days. As will be shown in Section 13.2, the events are characterized by fluxes of opposite sign, i.e. the deposition of pollutants causes a release of CO_2 from the ecosystem. These findings may be interpreted as the outcome of stress conditions. Such phenomena have to date not been reported in the literature on forest decline field studies.

In Fig. 13.1a–d for the gases SO_2, NO, NO_2 and O_3 in each case two episodes with rather high concentrations are presented. As far as possible, events were selected in which essentially only one of the gases is involved in the interrelation. Very often superpositions occur, in particular in the case of the nitrogen oxides. For instance, the last CO_2 maximum in the left part of Fig. 13.1b is due to the effect of nitrogen monoxide. In Fig. 13.1c, in addition to the nitrogen monoxide, NO_2 also contributes to the alteration of the CO_2 concentration. On the one hand, this gas induces additional maxima, on the other hand, it gives rise to an increase in the CO_2 concentration before the true effect by the monoxide occurs. In the left part of Fig. 13.1a, the nitrogen oxides also contribute to the response of the ecosystem during the day after the SO_2 maximum. On the whole the results of the gas measurements seem to suggest complex interrelations between sulphur dioxide, the nitrogen compounds and carbon dioxide. The peak values in the case of SO_2 and O_3 amounted to as high as 720 $\mu g/m^3$ and 330 $\mu g/m^3$, respectively. Concentrations in this order of magnitude with an impact of several hours or

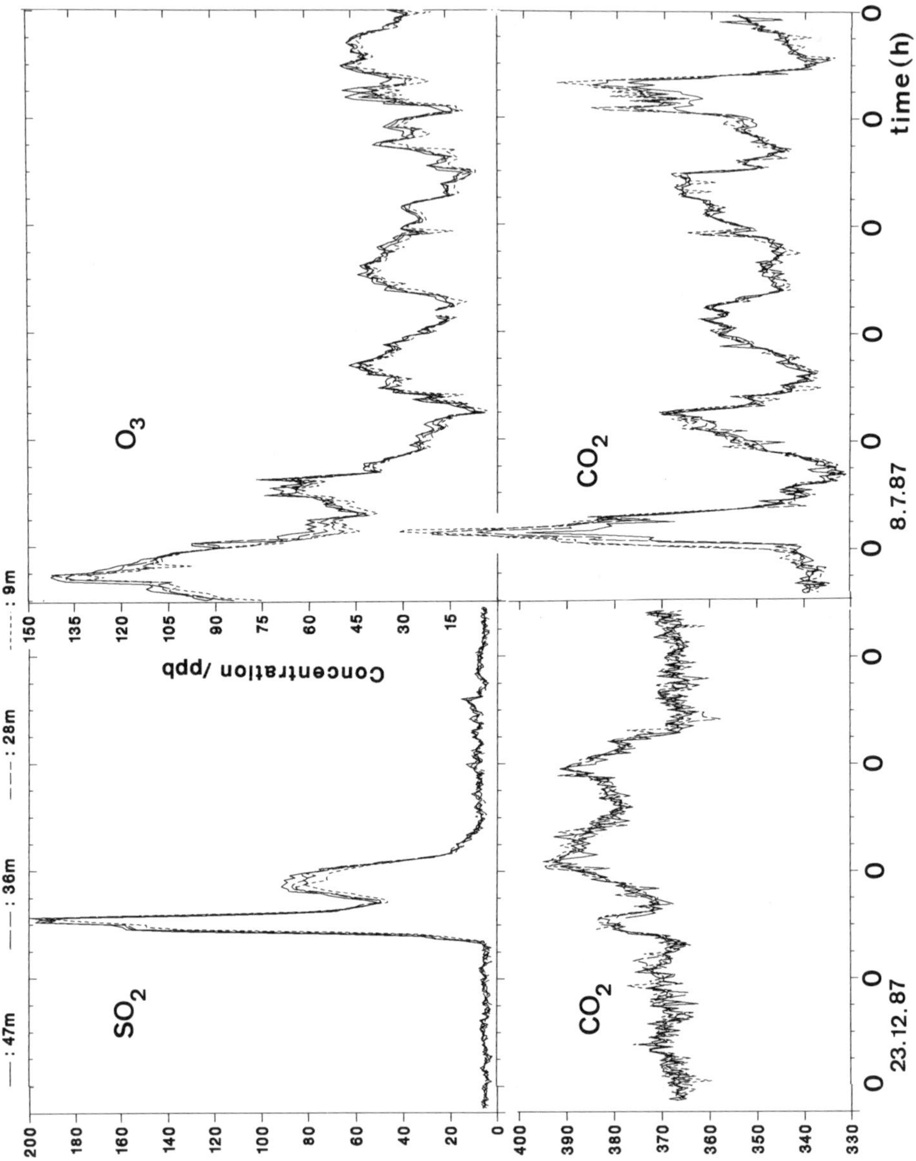

Fig. 13.1a–d. Interrelations between the concentrations of gaseous pollutants and carbon dioxide

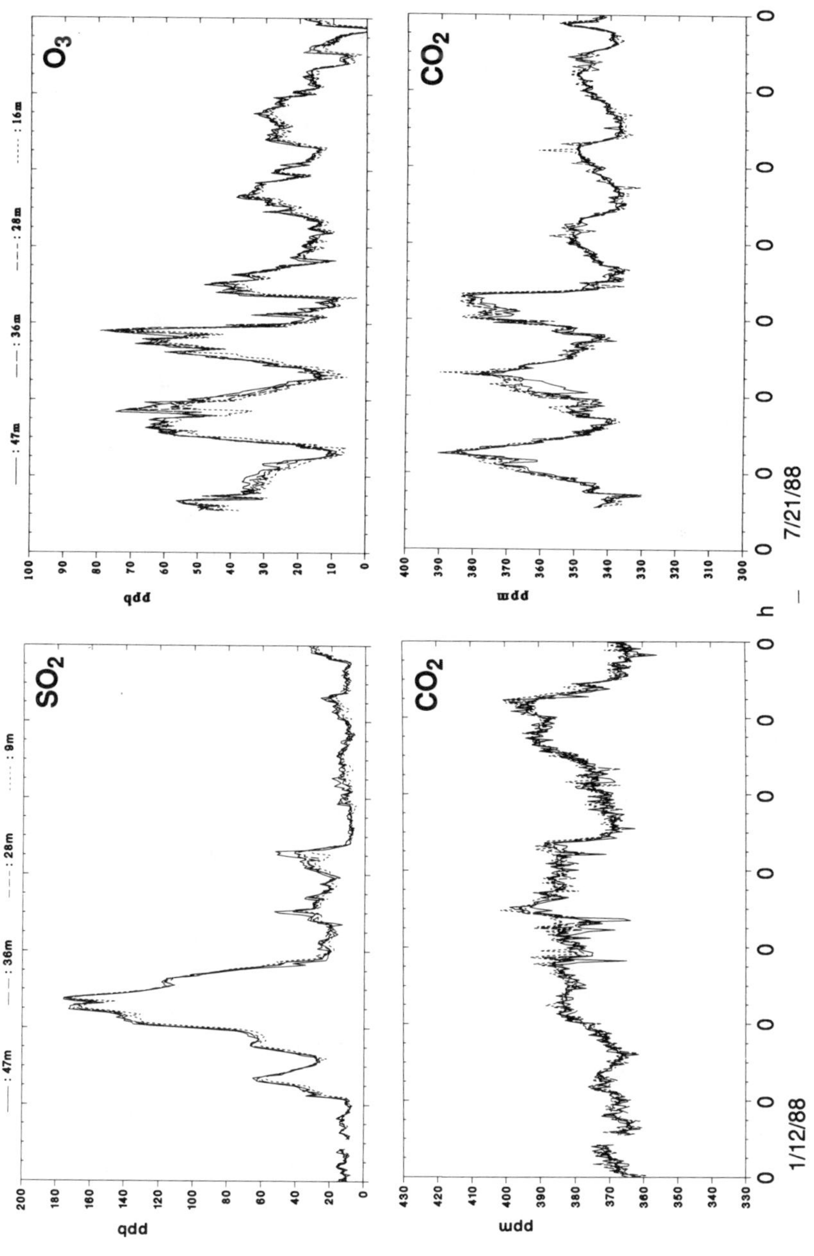
SO2
CO2
O3
CO2
: 47m
: 36m
: 28m
: 9m
: 16m
ppb
ppm
h
1/12/88
7/21/88

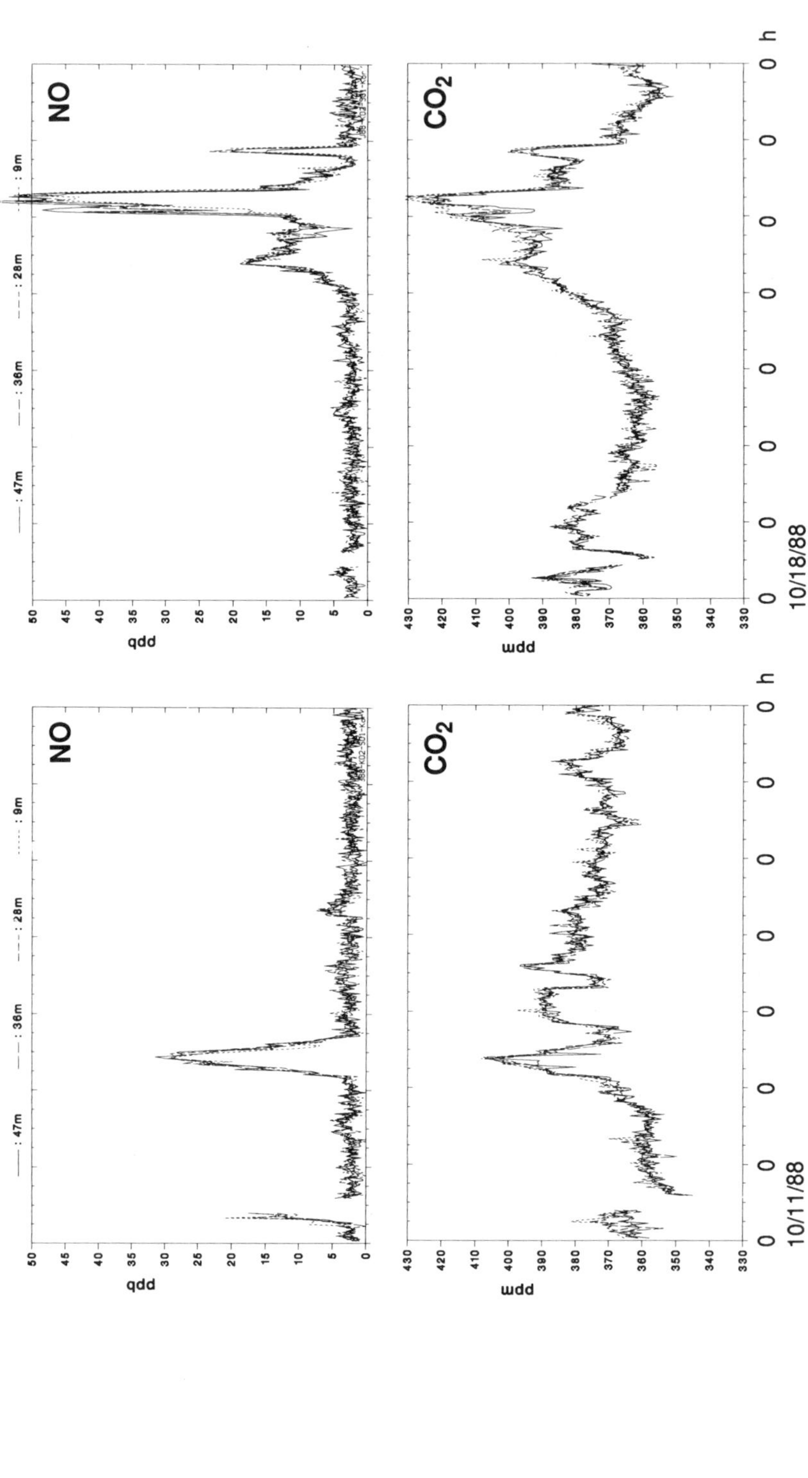

—— : 47m
—— : 36m
- - - : 28m
...... : 9m
NO
CO$_2$
ppb
ppm
10/18/88
10/11/88
0 h

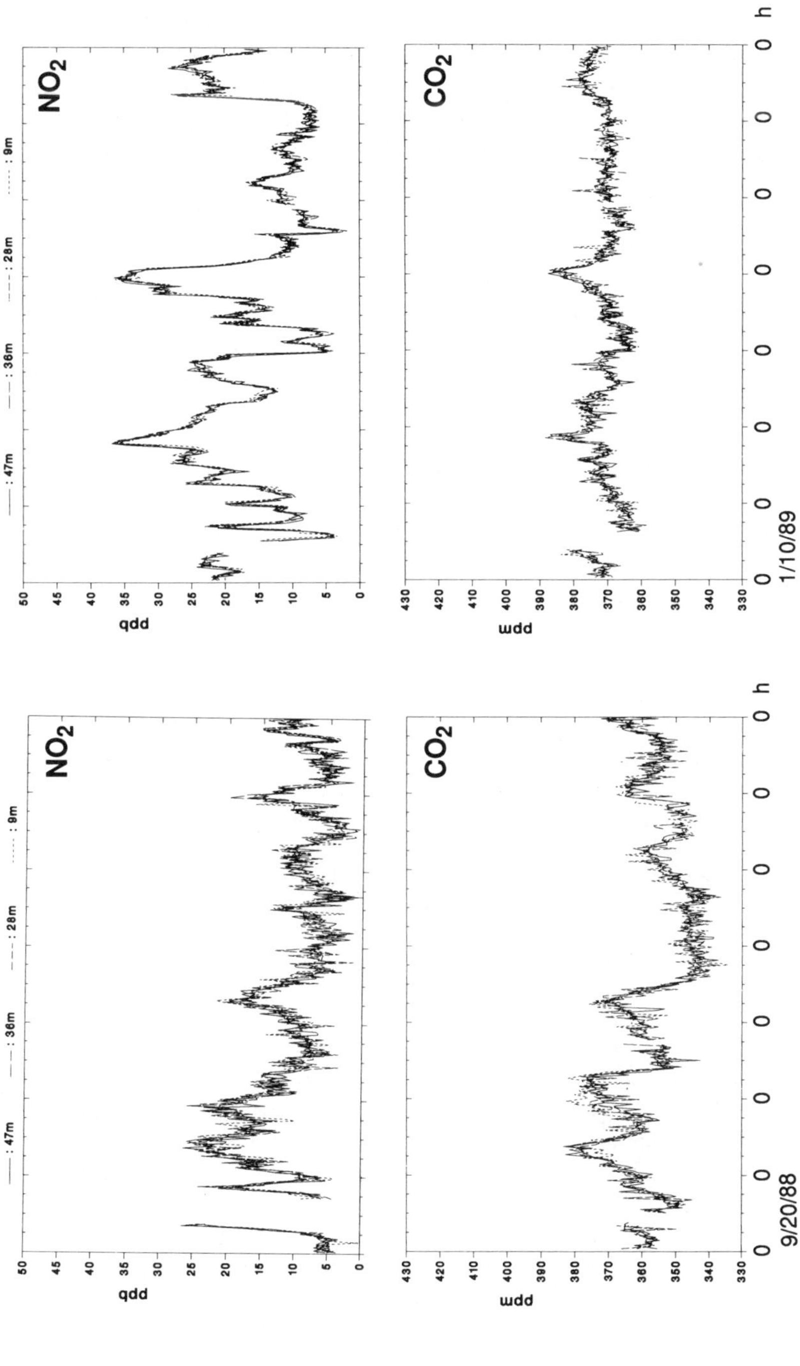

NO2
CO2
ppb
ppm
: 47m
: 36m
: 28m
: 9m
9/20/88
1/10/89
h

days in all probability clearly exceed the damage producing concentration limits (Keller 1976; Arndt et al. 1982; Lichtenthaler 1984).

The result of the present study that the fluxes of gaseous pollutants and carbon dioxide have opposite sign is of great importance for the interpretation of the observed phenomena. The different gradients already become obvious from Fig. 13.1. More details will be given in Section 13.2, in particular with respect to the most significant gases SO_2 and O_3. In the case of these two pollutants, the temporal behaviour of the response of the forest ecosystem is quite different. While a rise in the concentration of SO_2 induces an instantaneous reaction, the results in the case of O_3 indicate a correlation between the ozone maximum by day and the carbon dioxide maximum values in the early morning of the next day (Fig. 13.1). The response of the ecosystem appears to intensify after repeated high immissions. The time-shift of the response in the case of O_3 points out that the mechanism of action is different from that of SO_2 (cf. Sect. 13.2).

The nitrogen oxides show a temporal behaviour which is similar to the findings obtained for sulphur dioxide. However, it should be stressed that the deposition events of these gases are continuously attended by episodes with release into the atmosphere. This is exemplified in Fig. 13.2 by means of the gradients in ppb/m (reference height 14 m above canopy) during September 1988. Hence, it appears that the sign changes continuously. The plots reproduce the measurements with the original time resolution of 30 min. Figure 13.2 accounts for the quasi-vanishing annual mean values of the turbulent fluxes of NO and NO_2 summarized in Table 12.2. It may be concluded that these gases alone do not cause severe damage to the forest ecosystem (cf. World Health Organization 1985; Sect. 12.2). However, the combined impact together with SO_2 is obviously quite harmful. This conclusion is confirmed by another study within the "Postturm" project (Lalk et al. 1992). In growth chamber experiments it was shown that even rather low concentrations of SO_2 and NO_2 result in a depression of net photosynthesis and transpiration as well as disturbances in the chlorophyll fluorescence.

The findings concerning the interrelations between the concentrations of gaseous pollutants and carbon dioxide are valid for the total forest ecosystem including all compartments. However, the gas analysis at a height of 1 m above the ground indicates at least perceptible contributions of the soil to the release of carbon dioxide. This becomes obvious from Figs. 13.3 and 13.4. In the first case, the concentrations of SO_2 and O_3 measured at 9, 28, 36 and 47 m are plotted together with the concentrations of CO_2 at 1, 9, 28 and 47 m during a period of three successive days in October 1990. This event was characterized by rather stable meteorological conditions with winds from southeast (5 to 6 m/s at 47 m), an absence of rainfall and an air pressure of about 1010 mbar. The values of global radiation, relative humidity and temperature at 2 p.m. amounted to 350 to 375 W/m^2, 68 to 75% and 8 to 9 °C, respectively. During each day, clear maxima of SO_2 and O_3 occurred. As

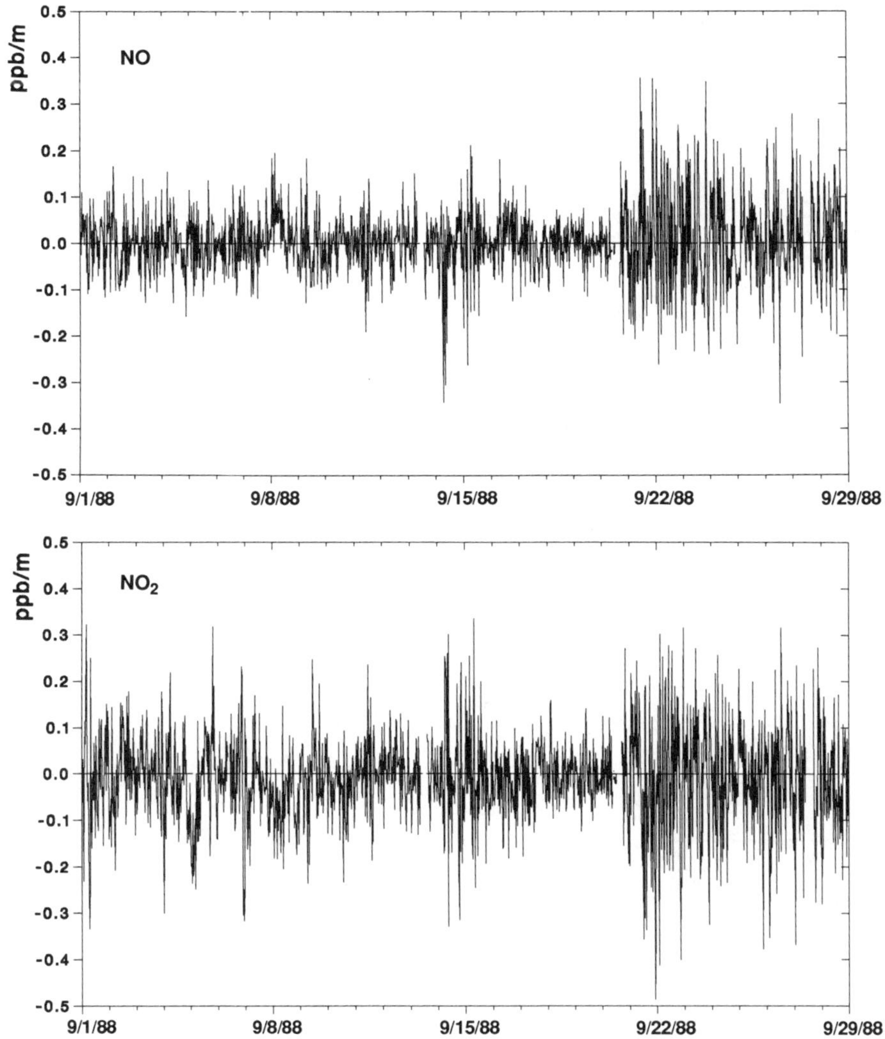

Fig. 13.2. Temporal variation in the concentration gradients of NO and NO_2 in ppb/m during September 1988

shown in Fig. 13.3, nearly throughout the whole 3-day period the CO_2 concentration at 1 m height aboveground definitely exceeded the values at the other measurement heights. This result clearly indicates the enhanced release of CO_2 from the soil. Quantitative statements with regard to the corresponding fluxes, however, require additional equipment in order to determine the correct exchange coefficients between soil and crown compartment. Decomposition processes in the L, O_{fh} and A_h horizons are most probably the sources of the carbon dioxide originating from the soil (cf.

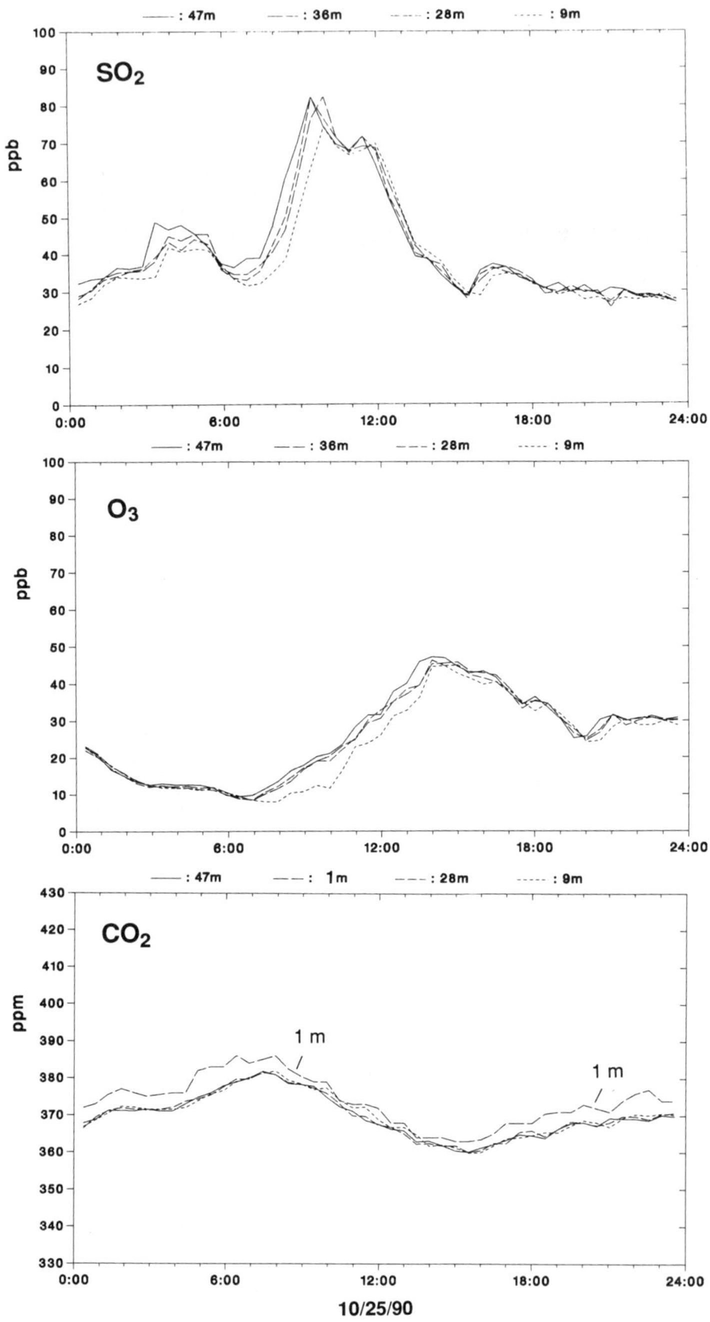

Fig. 13.3a–c. Concentrations of SO_2 and O_3 measured at 9, 28, 36 and 47 m together with the concentrations of CO_2 at 1, 9, 28 and 47 m during a period of three successive days (25–27 October 1990)

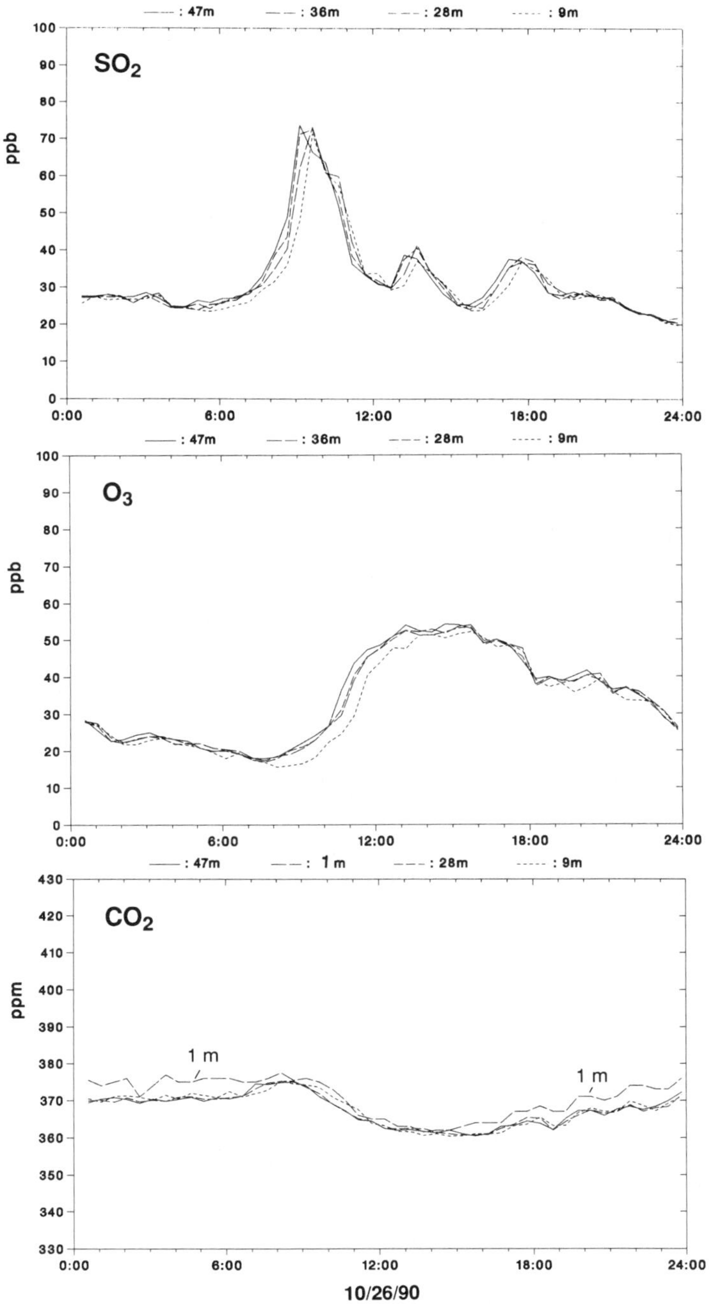
: 47m
: 36m
: 28m
: 9m
SO2
ppb
0:00
6:00
12:00
18:00
24:00
O3
ppb
: 1 m
CO2
ppm
1 m
1 m
10/26/90

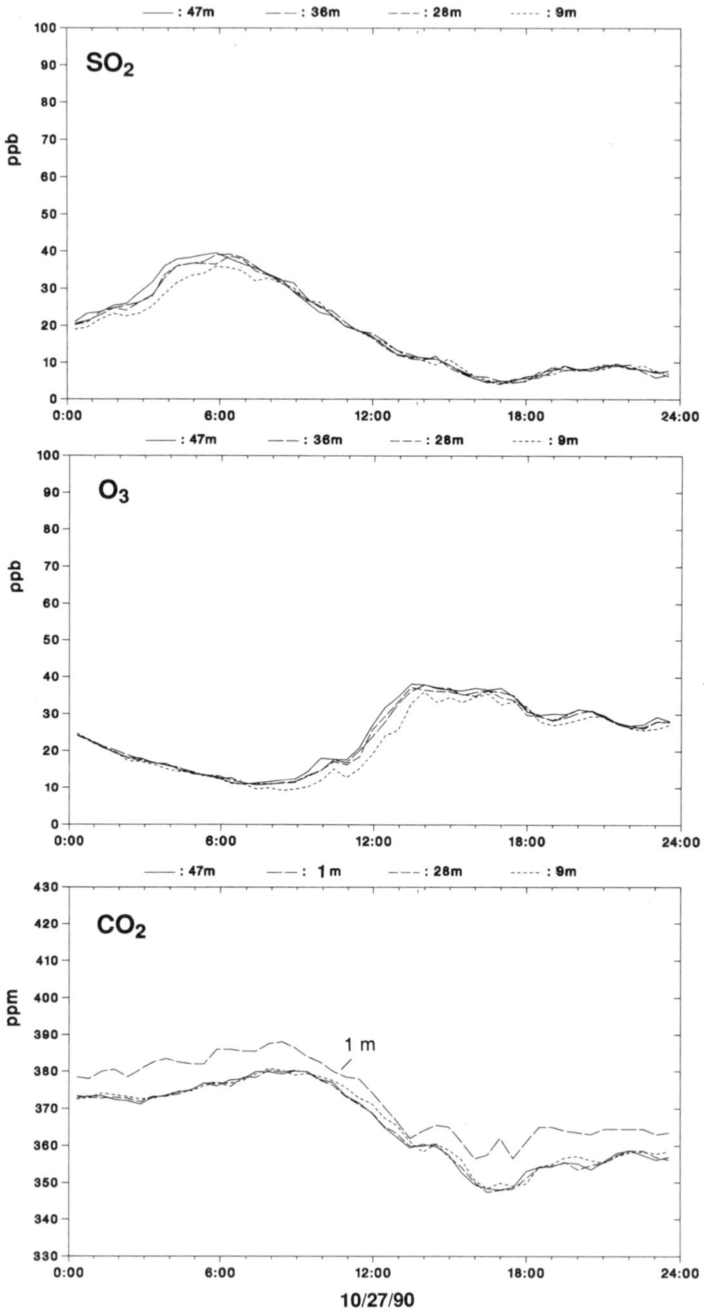

—— : 47m
— — : 36m
- - - : 28m
····· : 9m
SO2
ppb
100
90
80
70
60
50
40
30
20
10
0
0:00
6:00
12:00
18:00
24:00
—— : 47m
— — : 36m
- - - : 28m
····· : 9m
O3
ppb
100
90
80
70
60
50
40
30
20
10
0
0:00
6:00
12:00
18:00
24:00
—— : 47m
— — : 1 m
- - - : 28m
····· : 9m
CO2
ppm
430
420
410
400
390
380
370
360
350
340
330
1 m
0:00
6:00
12:00
18:00
24:00
10/27/90

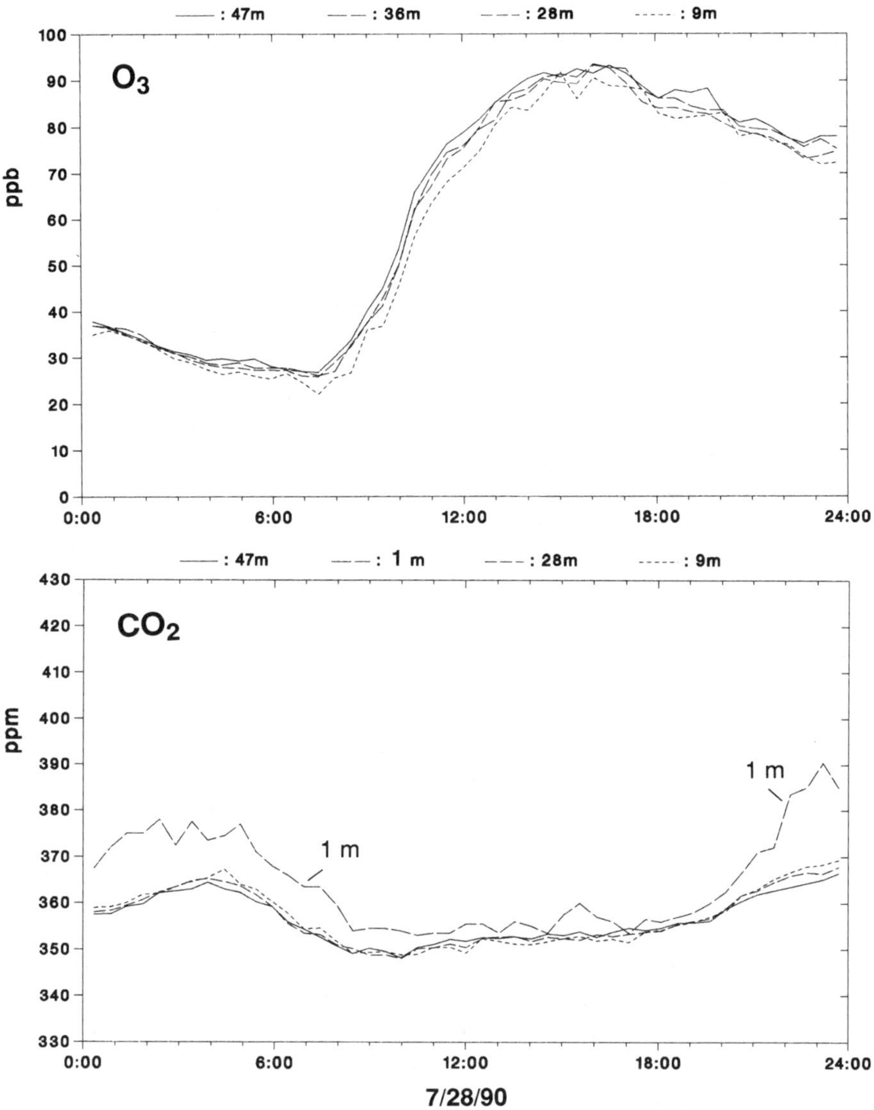

Fig. 13.4. Diurnal variation in the concentrations of O_3 (9, 28, 36 and 47 m) and CO_2 (1, 9, 28 and 47 m) during a summer day

Sect. 13.2). Details of the mechanisms under pollutant impact and the reasons for the different temporal courses have not as yet been completely clarified.

Another example of the reaction of the soil is presented in Fig. 13.4. In this case, the conditions were almost exclusively governed by high O_3 concentrations during a typical summer episode (July 1990). The nocturnal enhancement of the CO_2 concentration near the ground is very pronounced, the high values in the early morning being due to a similar impact of O_3 during the preceding day.

13.2 Associated Vertical Fluxes

As has already been demonstrated in Section 13.1, the fluxes of gaseous pollutants and carbon dioxide have opposite signs. This statement is clearly evidenced in Fig. 13.5 which shows two examples for the course of the gradients during two stress periods involving SO_2 and O_3 impact, respectively (Michaelis et al. 1989a,b). While in the first case the interpretation is quite straightforward on the basis of the results presented above, the temporal behaviour in the second case is rather complex and appears to contradict the conclusions made in Section 13.1. The reason lies in the fact that all the first 4 days of the period shown were characterized by both daily and distinct nocturnal ozone maxima. The course of the concentration during the third day was presented in Fig. 12.19a. During the night, the gradients are particularly pronounced.

For a few episodes the CO_2 gradients have been plotted in Fig. 13.6 as a function of the gradients of SO_2 and O_3, respectively (Michaelis et al. 1989a,b). In the case of O_3 the events during the vegetation period and during the vegetation rest are distinguished by full and open circles. Two conclusions may be drawn from Fig. 13.6. Firstly, the effect of O_3 is more severe than that of SO_2. Secondly, the impact of O_3 does not depend very strongly on the season. One might speculate that this perhaps supports the hypothesis of a marked contribution from the soil to the CO_2 balance during pollutant impacts. The fluxes observed for SO_2 range between 0 and -5 μg/m²s, those of O_3 range between 0 and -6 μg/m²s. Depending on the kind of pollutant, the associated release of CO_2 amounts to as much as 5 mg/m²s and 10 mg/m²s, respectively. In Fig. 13.7, the correlation between the fluxes of SO_2 and CO_2 for a number of events is indicated (Michaelis et al. 1991). Of course, due to the almost permanent presence also of other pollutants in plots such as those given in Figs. 13.6 and 13.7 the number of usable events is rather limited, and a complete absence of interference effects can hardly be assured. The error limits are therefore fairly high. In the case of the fluxes of ozone and carbon dioxide, it is more convenient to examine long-term values.

Since half-hour values over a long period are available, further conclusions can be drawn with the aid of a computer simulation in which events with a high ozone impact above given threshold values are eliminated (Michaelis and Theopold 1993). In this procedure two O_3 threshold values were chosen: I deposition ≤ -0.43 μg/m²s or concentration ≥ 40 ppb; II deposition ≤ -0.65 μg/m²s or concentration ≥ 60 ppb. For such events in the

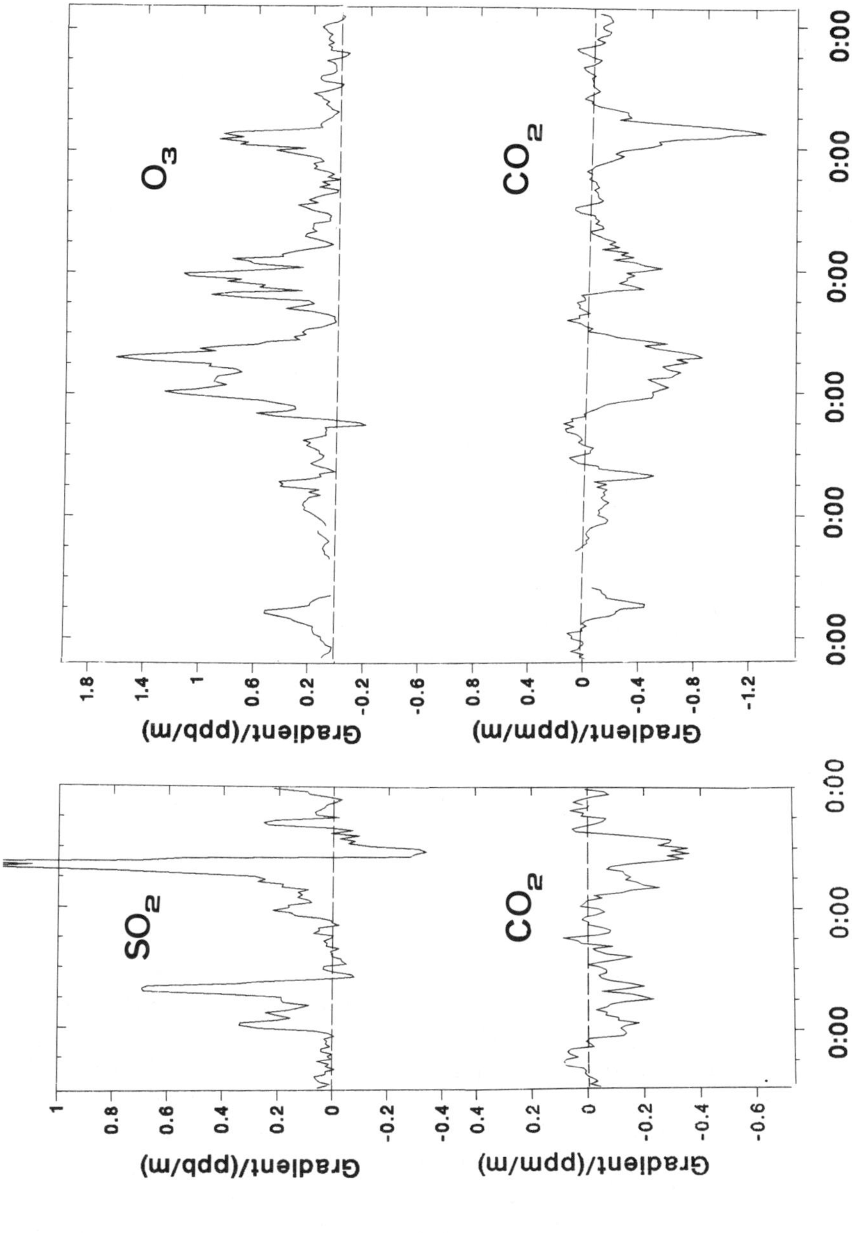

Fig. 13.5. Interrelations between gradients of SO_2/O_3 and CO_2

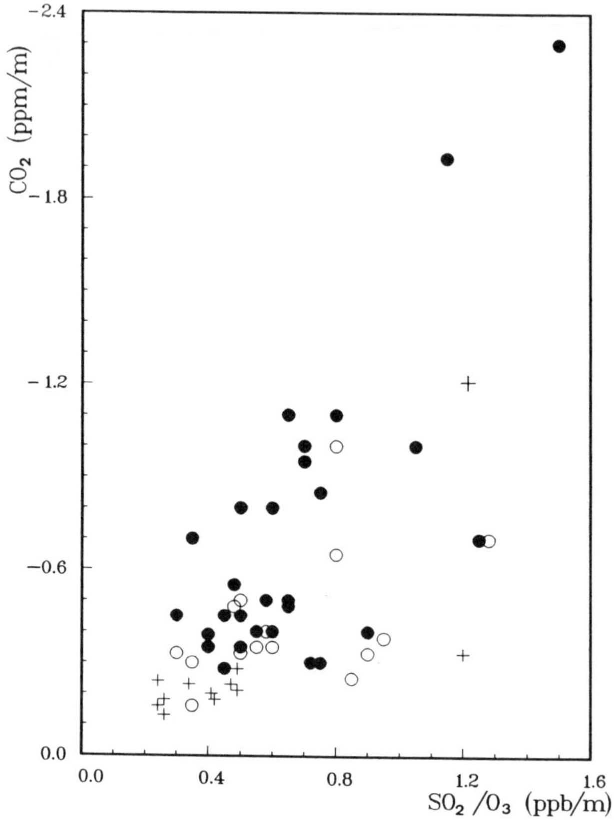

Fig. 13.6. Plot of CO_2 gradients vs. SO_2 (*crosses*) and O_3 (*circles*) gradients. *Full circles* During and *open circles* out of the vegetation period

CO_2 balance the corresponding data measured between 6 p.m. and 10 a.m. were eliminated and replaced by long-term mean values obtained under conditions without O_3 pollution. The result is presented in Table 13.1. Obviously, the turbulent fluxes of CO_2 improve towards negative values with decreasing ozone impact. This conclusion must be seen in context with a study performed in the early 1970s in a Bavarian forest stand (Hager 1975). Under environmental conditions which were much more favourable than in the present study, a CO_2 flux of about -0.1 mg/m^{2}s was measured. Thus ozone is certainly a gaseous pollutant which requires particular attention. A similar though minor effect was also observed in the case of SO_2. Nevertheless, the trend shown in Table 13.1 is further enhanced. In the case of the nitrogen oxides, no significant alterations were detected. With regard to all these findings it should be pointed out, however, that in view of the intricate determination of turbulent fluxes and the possible resulting experimental errors the data have a primarily qualitative character.

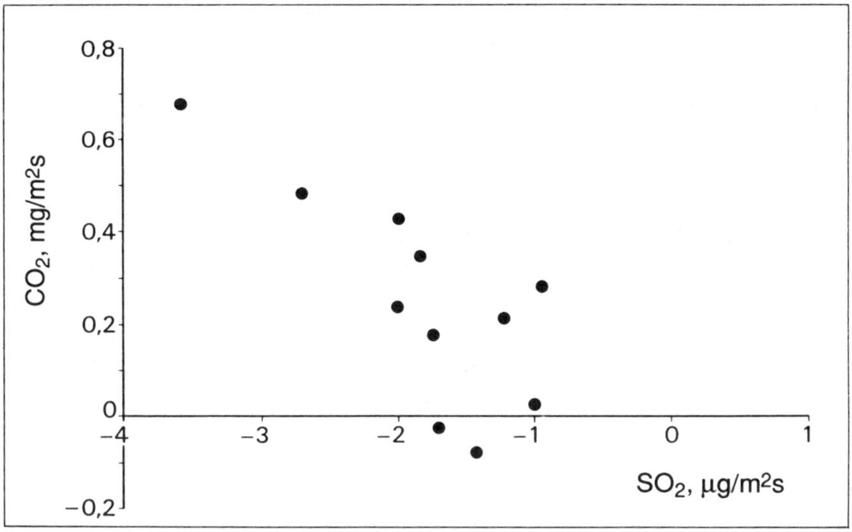

Fig. 13.7. Interrelation between the turbulent fluxes of SO_2 and CO_2. Each *point* represents the mean value of events with SO_2 concentrations within a 10 ppb interval. Lower threshold value 50 ppb

Table 13.1. Annual mean values of the turbulent fluxes of carbon dioxide including all events of O_3 impact (a) and after elimination of events exceeding two different threshold values (b, c). See text. All data in mg/m^2s

	1987	1988	1989	1990	1991	Weighted mean
a. Values from Table 12.2	0.032	0.062	- 0.002	0.026	0.022	0.028
b. Threshold value II	- 0.001	-	- 0.023	0.007	- 0.007	- 0.012
c. Threshold value I	0.033	-	- 0.092	0.000	- 0.004	- 0.016

All these results have to be seen in the context of numerous plant-physiological studies. It is not the object of this volume to attempt an interpretation on the basis of physiological fundamentals, the less so as further detailed experiments are certainly necessary, particularly with regard to the fluxes inside the forest stand. However, references to some probably relevant studies are without doubt useful. Several publications have dealt with the flux of SO_2 into leaves, the mesophyll resistances to these fluxes, the cellular acidification by SO_2 as well as the photosynthetic capacity and carboxylation efficiency of Norway spruce trees (Lange et al. 1985, 1986a,b, 1987, 1989a,b; Lange and Zeller 1986; Pfanz et al. 1987a,b; Oren and Zimmermann 1989). These studies, which were mainly performed in the Fichtelgebirge (FRG), suggest that direct effects of atmospheric pollutants weaken the trees and make them susceptible to impairment caused by soil acidification and mineral deficiencies. Damaged trees showed a strong depression of photosynthetic capacity and carboxylation efficiency. These findings were connected with significantly reduced concentrations of the element Mg in chlorotic needles. Diminished contents were also observed in the case of Ca, Mn and Zn, whereas the concentrations of K and Al were markedly increased. When comparing these results with the data presented in Fig. 11.12 and Table 11.5 for the "Postturm" site, a transferability can hardly be diagnosed. Moreover, the different experimental conditions have to be taken into account.

Fumigation experiments with O_3, SO_2 and the combination O_3/SO_2 in greenhouses and climatic chambers (Vogels et al. 1986; Küppers and Klumpp 1988) also indicated direct effects of atmospheric pollutants. Photosynthesis in 4-year-old Norway spruce clones was reduced, whereas dark respiration was stimulated. Ozone and the combination O_3/SO_2 were most effective, causing an increase in respiration of up to 40% in the case of current-year needles relative to the control. The experiments were performed with exposure times in the order of several months and concentrations between 50 and 200 $\mu g/m^3$ ozone and 75 to 100 $\mu g/m^3$ sulphur dioxide. These concentration levels are comparable with those during the stress episodes presented in Figs. 13.1 and 13.3. The time resolution, however, does not allow any direct conclusions to be made on the short-term measurements at the "Postturm" site. A study of the influence of SO_2 fumigation under field conditions on the metabolism of beech leaves (*Fagus sylvatica* L.) by means of ^{14}C labelling supports the above-mentioned results for Norway spruce (Landolt 1982). Within a period of 5 weeks the uptake of CO_2 decreased to about 50% compared to that of the control.

Short-term fumigation experiments with SO_2 and O_3 were performed by Saxe and Murali (1989a,b) as well as within the "Postturm" project by Lalk et al. (1992). In the first of these studies, a 4-h impact of 500 ppb SO_2 on 4-year-old spruce clones caused a reversible decrease in the net photosynthesis and the transpiration. At concentration levels above 1280 ppb the effects were no more completely reversible. Ozone concentrations higher than 166 ppb induced irreversible reductions of the gas exchange. Lalk et al. (1992) observed during a 31-h exposure with about 830 ppb SO_2 a decrease in the net photosynthesis by nearly 50% and alterations in the chlorophyll fluorescence. The experiments were performed with 7- to 10-year-old spruce clones in climatic chambers. In the case of O_3 a 30-h fumigation with 250 ppb caused a drastic drop in the net photosynthesis by about a factor of 4 as well as a distinct reduction of the transpiration. Obviously, there was a clear dependence of these effects on the physiological course of development of the needles.

As concerns the second possible source of CO_2, i.e. the forest soil (cf. Figs. 13.3 and 13.4), reference to several studies in the field of soil research might be very helpful (Brumme and Beese 1992; Loftfield et al. 1992;

Brumme 1995; Brumme and Beese 1995). The authors have developed an automated monitoring device for the measurement of trace gas fluxes from forest soils, they determined the fluxes of CO_2 and NO_2 under different conditions and investigated the mechanisms of carbon and nutrient release and retention. It would be very interesting to extend such experiments under controlled impact of SO_2 and O_3.

References

Arndt U, Seufert G, Nobel W (1982) Die Beteiligung von Ozon an der Komplexkrankheit der Tanne (*Abies alba* Mill.) - eine prüfenswerte Hypothese. Staub-Reinh Luft 4(2),6:243–247

Brumme R (1995) Mechanisms of carbon and nutrient release and retention in beech forest gaps. Plant Soil 168–169:593–600

Brumme R, Beese F (1992) Effects of liming and nitrogen fertilization on emissions of CO_2 and N_2O from a temporate forest. J Geophys Res 97, D12:12 851–12 858

Brumme R, Beese F (1995) Automated monitoring of biological trace gas production and consumption. In: Alef K, Mannipieri P (eds) Methods in applied soil microbiology and biochemistry. Academic Press, London, pp 468–472

Hager H (1975) Kohlendioxid - Konzentrationen, Flüsse und Bilanzen in einem Fichtenhochwald. Münchner Universitäts-Schriften, Fachbereich Physik, Nr. 26

Keller T (1976) Auswirkungen niedriger SO_2-Konzentrationen auf junge Fichten. Schweiz Z Forstwesen 127:237–251

Küppers K, Klumpp G (1988) Effects of ozone, sulfur dioxide, and nitrogen dioxide on gas exchange and starch economy in Norway spruce (*Picea abies* (L.) Karst.). GeoJournal 17(2):271–275

Lalk I, Hartmann A, Dörffling K (1992) Wirkung kurzzeitiger Schadgas-Expositionen (SO_2, NO_2, O_3) auf geklonte Jungfichten im Simulationsexperiment. In: Michaelis W, Bauch J (eds) Luftverunreinigungen und Waldschäden am Standort "Postturm", Forstamt Farchau/Ratzeburg. GKSS Forschungszentrum Geesthacht, GKSS 92/E/100, pp 309–340

Landolt W (1982) Der Einfluß einer praxisnahen SO_2-Begasung auf das ^{14}C-Fixierungsmuster von Buchen (*Fagus silvatica* L.). Eur J For Pathol 12:331–339

Lange OL, Zellner H (1986) Physiologische Veränderungen bei geschädigten Bäumen. In: Führ F, Ganser S, Kloster G, Prinz B, Stüttgen E (eds) Statusseminar 1985 der Arbeitsgruppe "Waldschäden/Luftverunreinigungen", Kernforschungsanlage Jülich, pp 326–338

Lange OL, Gebel J, Schulze ED, Walz H (1985) Eine Methode zur raschen Charakterisierung der photosynthetischen Leistungsfähigkeit von Bäumen unter Freilandbedingungen - Anwendung zur Analyse "neuartiger Waldschäden" bei der Fichte. Forstwiss Centralbl 104:186–198

Lange OL, Führer G, Gebel J (1986a) Rapid field determination of photosynthetic capacity of spruce twigs (*Picea abies*) at saturating ambient CO_2. Trees 1:70–77

Lange OL, Gebel J, Zellner H, Schramel P (1986b) Photosynthesekapazität und Magnesiumgehalte verschiedener Nadeljahrgänge bei der Fichte in Waldschadensgebieten des Fichtelgebirges. In: Führ F, Ganser S, Kloster G, Prinz B, Stüttgen E (eds) Statusseminar 1985 der Arbeitsgruppe "Waldschäden/Luftverunreinigungen", Kernforschungsanlage Jülich, pp 127–147

Lange OL, Zellner H, Gebel J, Schramel P, Köstner B, Czygan FC (1987) Photosynthetic capacity, chloroplast pigments, and mineral content of the previous year's spruce needles with and without the new flush: analysis of the forest-decline phenomenon of needle bleaching. Oecologia 73:351–357

Lange OL, Weikert RM, Wedler M, Gebel J, Heber U (1989a) Photosynthese und Nährstoffversorgung von Fichten aus einem Waldschadensgebiet auf basenarmen Untergrund. Allg Forst Z 3/1989:55–64

Lange OL, Heber U, Schulze ED, Ziegler H (1989b) Atmospheric pollutants and plant metabolism. In: Schulze ED, Lange OL, Oren R (eds) Forest decline and air pollution. Ecological Studies 77. Springer, Berlin Heidelberg New York, pp 238–273

Lichtenthaler HK (1984) Luftschadstoffe als Auslöser des Baumsterbens. Naturwiss Rundsch 37(7):271–277

Loftfield NS, Brumme R, Beese F (1992) Automated monitoring of nitrous oxide and carbon dioxide flux from forest soils. Soil Sci Soc Am J 56:1147–1150

Michaelis W, Theopold F (1993) Deposition atmosphärischen Ozons und ihre Wirkung auf ein Waldökosystem. In: Arbeitsgemeinschaft der Großforschungseinrichtungen (AGF)(ed) Atmosphärisches Ozon. Prozesse und Wirkungen. Thenée Druck, Bonn, pp 25–27

Michaelis W, Schönburg M, Stößel RP (1988) Trocken- und Naßdeposition von Schwermetallen und Gasen. In: Bauch J, Michaelis W (eds) Das Forschungsprogramm Waldschäden am Standort "Postturm", Forstamt Farchau/Ratzeburg. GKSS Forschungszentrum Geesthacht, GKSS 88/E/55, pp 19–59

Michaelis W, Schönburg M, Stößel RP (1989a) Deposition of atmospheric pollutants into a North German forest ecosystem. In: Georgii HW(ed) Mechanisms and effects of pollutant-transfer into forests. Kluwer, Dordrecht, pp 3–12

Michaelis W, Schönburg M, Stößel RP (1989b) Schadstofftransfer in der Grenzschicht Atmosphäre-Vegetation. In: Arbeitsgemeinschaft der Großforschungseinrichtungen (AGF) (ed) Wechselwirkung Atmosphäre-Biosphäre. Thenée Druck, Bonn, pp 29–33

Michaelis W, Pepelnik R, Rademacher P, Riebesell M (1990) Wechselwirkung zwischen Luftschadstoffen und Vegetation. In: GKSS Jahresbericht 1990, GKSS Forschungszentrum Geesthacht, pp 42–55

Michaelis W, Pepelnik R, Rademacher P, Riebesell M (1991) Transfer of atmospheric pollutants into a forest ecosystem. In: Teller A, Mathy P, Jeffers JNR (eds) Responses of forest ecosystems to environmental changes. Elsevier, London, pp 596–597

Michaelis W, Pepelnik R, Theopold F, Rademacher P (1992) Deposition atmosphärischer Spurenstoffe und Stoffflüsse im Ökosystem Wald. In: Michaelis W, Bauch J (eds) Luftverunreinigungen und Waldschäden am Standort "Postturm", Forstamt Farchau/Ratzeburg. GKSS Forschungszentrum Geesthacht, GKSS 92/E/100, pp 11–59

Oren R, Zimmermann R (1989) CO_2 assimilation and the carbon balance of healthy and declining Norway spruce stands. In: Schulze ED, Lange OL, Oren R (eds) Forest decline and air pollution. Ecological Studies 77. Springer, Berlin Heidelberg New York, pp 352–369

Pfanz H, Martinoia E, Lange OL, Heber U (1987a) Mesophyll resistances to SO_2 fluxes into leaves. Plant Physiol 85:922–927

Pfanz H, Martinoia E, Lange OL, Heber U (1987b) Flux of SO_2 into leaf cells and cellular acidification by SO_2. Plant Physiol 85:928–933

Saxe H, Murali NS (1989a) Diagnostic parameters for selecting against novel spruce (*Picea abies*) decline. I. Tree morphology and photosynthesis response to acute SO_2 exposures. Physiol Plant 76:340–348

Saxe H, Murali NS (1989b) Diagnostic parameters for selecting against novel spruce (*Picea abies*) decline. III. Response of photosynthesis and transpiration to O_3 exposures. Physiol Plant 76:356–361

Vogels K, Guderian R, Masuch G (1986) Studies on Norway spruce (*Picea abies* Karst.) in damaged forest stands and in climatic chamber experiments. In: Schneider T (ed) Acidification and its policy implications. Studies in Environmental Science 30. Elsevier, Amsterdam, pp 171–186

World Health Organization (1985) Air quality guidelines – ecological effects of air pollutants. ICP/CEH 902/m 71 (S), Geneva

14 Summary and Conclusions

The present volume gives a comprehensive résumé of the research work carried out by the GKSS Research Centre Geesthacht at a forest site in North Germany during the period 1986 to 1992. The study was performed within the framework of the so-called Postturm project. Central objectives were to generate extensive data sets on air quality including trends and pollutant source-receptor relationships, to investigate the interaction of atmospheric constituents with the forest ecosystem compartments and thus to provide a reliable basis for the assessment of possible impacts on the state of health of the forest.

Such an ambitious goal cannot be reached by simple immission measurements alone. For this reason special features of the investigation were additional measurements of the deposition of atmospheric trace substances into the forest and the determination of the associated fluxes within the ecosystem including their balances. This procedure makes large demands on the accuracy and precision of the analytical data as well as on the detection sensitivity. Preconditions are reliable sampling techniques and high-performance analytical methods. In the case of trace elements, total-reflection X-ray fluorescence (TXRF) and inductively coupled plasma optical emission spectroscopy (ICP-OES) have proved to be very successful. Systematic intercomparisons with other analytical techniques also supported the quality assurance. Gas analysis was performed using UV and IR absorption, UV fluorescence and chemiluminescence. Here, particular attention was paid to the properties of the air sampling pipe system in order to exclude systematic errors in the determination of the concentration gradients.

Other important sources of errors are the methodologies underlying the derivation of fluxes from the analytical data. This is particularly relevant in the case of the 'concentration method' which is usually applied for determining the dry deposition of trace elements via airborne particulates. Besides correct size fractionation, the central problem of this method lies in the state of knowledge of the deposition velocity which strongly depends on the equivalent aerodynamic diameter and the particle surface properties. During the initial stage of the project, severe discrepancies existed in the literature with regard to this subject. Therefore an attempt was made to solve the problem by combining the 'concentration' and the 'gradient method'. It turned out that the results of experiments which used artificial aerosols for the determination of the deposition velocity do not reflect the natural conditions. This conclusion is also relevant with respect to the impactor characteristics. The consistency which was achieved in the flux balances corroborates the correctness of the chosen procedure. Taking into account the different cases of atmospheric stratification, the gradient method was also successfully applied to the determination of the vertical fluxes of gaseous pollutants.

As many as 27 trace elements were determined in rainwater over a period of about 5 years. Dissolved and particulate phases were analyzed separately. The element pattern includes both toxic heavy metals and nutrients. Consideration of the latter elements is very important since the impact of atmospheric constituents, in particular hydrogen ions, severely affects the budget of the nutritional elements. Concentrations and wet deposition of pollutants show a distinct dependence on the wind direction. On the one hand, there is a strong influence of the conurbation of Hamburg which is situated about 40 km west-southwest of the investigation site. On the other hand, in the case of sulphur, additional high emissions in the distant browncoal mining and industrial area of the former German Democratic Republic markedly contribute to the air pollution. The annual mean values of the total wet deposition of several anthropogenic pollutants show a clearly decreasing tendency. This is particularly evident for the elements sulphur, arsenic, cadmium and lead. The trends substantiate the success of environmental policy measures. With the exception of sulphate the deposition of anions and cations via precipitation does not reveal any significant tendency. Particularly in the case of the essential H^+ ions no long-term trend became apparent during the 5-year investigation period.

Airborne particulates were analyzed separately in six size fractions with a detection sensitivity down to the order of pg/m^3. As many as 27 trace elements were detected. Again, the concentrations and the deposition exhibit a clear dependence on the meteorological conditions, although differences in behaviour between dry and wet deposition are observed since the frequency distribution of the wind direction shows less anisotropy than the distribution of the amount of precipitation. This meteorological phenomenon has some important consequences. Firstly, the geographic distributions of the sources are not identical for wet and dry deposition. Emissions southeast of the measuring station contribute more strongly to the dry deposition process. This resulted in a less pronounced trend during the first half of the investigation period due to the inferior environmental policy in the former German Democratic Republic. In the case of sulphur and lead, the effect is particularly evident. A further consequence is that these reduced tendencies are also transmitted to the total deposition of the trace elements since the high filter capability of a forest stand induces a marked deposition velocity and thus a prevailing contribution to the total flux. Analysis of rain and mist alone is therefore an insufficient basis for comprehensive studies of forest ecosystems. The acidification of the soil also gives rise to the mobilization and availability of constituents of deposited particulates. Hence the dry deposition can contribute substantially, for instance, to the high heavy metal concentrations in the soil solutions and in the fine roots of the trees. The relative contributions of wet and dry deposition to the total flux show distinct seasonal variations with maxima in the dry deposition during the winter period and minima in the second or third quarter of the year.

Further focal points of the trace analytical investigations at the "Postturm" site were the disclosure of the relations between the element concentrations in the soil solutions of diverse horizons and those in the tree compartments, as well as the determination of element fluxes which are important for the supply to the ecosystem. In these studies, seasonal variations of the element concentrations and collectives of both healthy and declining trees were also examined. During a 4-year series of measurements, the macronutrient concentrations in the soil solution of the O_{fh}/A-horizon were observed to decrease. This was found to be associated with rather high heavy metal concentrations. Aluminum is only of limited importance in this horizon. In deeper layers, however, there is a drastic increase in the concentration of Al in the soil solution and, as a consequence, in the fine roots. On the other hand, the uptake of calcium is obviously impeded. The percentage of the total effective cation-exchange capacity for cations of strong acids reaches values of about 90% already at a soil depth of approximately 5 cm. The pH values in the humus layer range from 2.7 to 3.5, in the upper mineral soil they are about 2.9 and in the deeper horizons they range from 3.2 to about 4.0. Thus these strata have to be assigned to the Fe, Al/Fe and Al buffer regions, respectively. The nutritional elements are concentrated in the humus layer. These results indicate that the physiological conditions for the forest stand are quite poor. The development of an efficient fine root system is strongly impeded. Only about 20% of the total fine root mass was found below 50 cm soil depth.

The unfavourable conditions in the soil are passed on to the plant organs. For instance, the Al ion has to be considered as a competitive element which at low pH values displaces elements such as calcium from the exchange places at the fine-root cell wall. Another example is the uptake of heavy metals. Concentrations of lead as high as 300 ppm were observed in the fine roots of damaged trees. There is ample evidence that heavy metals are also a contributing factor in the decline of forests. This conclusion is supported by a recent study in which the concentrations of phytochelatins were investigated and correlated with the degree of decline (Gawel et al. 1996). These compounds are intracellular metal-binding peptides that act as specific indicators of metal stress (Grill et al. 1988; Schat and Kalff 1992). Increased concentrations of lead were also detected in the needles of damaged trees at the "Postturm" site. In this case it may be presumed that the metal is also taken up directly from the atmosphere. The supply of nutrients to the needles was on the whole adequate, but rather poor in the case of magnesium.

A detailed analysis of the element content, the biomass and the storage in the tree compartments was performed, in order to provide a useful support for a differentiated evaluation of the supply, the inventory and the loss of elements. Twenty-nine tree components were considered in this study. The element concentrations show quite different distribution patterns. Multiplication of the element contents by the respective biomass reveals the element stores in the various components. The data demonstrate that the total stores of magnesium in the biomass already exceed the total exchangeable stores in the mineral soil. Even if the stores bonded in the humus layer are also included in this stock-taking, the reservoir in the soil is only moderately higher than the amount fixed in the live biomass. These results may have grave consequences for the timber industry. Under unchanged environmental conditions, the next tree generation has to grow with a critical magnesium supply. The situation is even more severe in the case of potassium. As regards the element calcium, the data do not indicate a significant deficiency in the immediate future.

In all these considerations, the balances of the element fluxes also have to be taken into account. To this end, independent determinations of the fluxes (1) atmospheric deposition, (2) throughfall and (3) seepage water were performed. The balance total deposition minus seepage flux reveals that in the case of magnesium and calcium the export from the ecosystem is approximately three times higher than the input. At present, the removal of potassium is still compensated by a higher atmospheric deposition, though compared with the stores in the biomass and the yearly incorporation the effect is very low. The comparison of total deposition and throughfall reveals three groups of elements. The first group consists of those constituents which are retained in the crown compartment. Typical examples are hydrogen ions and heavy metals such as zinc and lead. The second group comprises elements which are characterized by higher concentrations in the throughfall. The most important examples are the nutritional elements magnesium, potassium, calcium and manganese. They are obviously leached out from the canopy and must be restored via the roots. In the case of potassium the throughfall exceeds the atmospheric input by more than a factor of 6, so that this element takes over the main part of the proton buffering in the canopy. Finally, there is a group of elements which show more or less neutral behaviour in the crown compartment. On the basis of all these findings concerning the nutritional elements, fertilization experiments were also performed within the scope of the "Postturm" project (Dünisch et al. 1992; Rademacher and Kriebitzsch 1992). Spruce plots were treated with potassium, magnesium and calcium. Compared to untreated control plots a wide range of positive effects could be achieved. These include improvement of biomass production, an increase in the nutrient contents in the roots, the phloem and the needles, an increase in needle weight, re-greening of discoloured needles and stabilization of the tree vitality during dry spells.

By analogy with the trace element measurements, the influence of the prevailing weather conditions is also evident in the immission behaviour of gaseous pollutants. This was particularly pronounced in the case of sulphur dioxide, the concentration of which was controlled to a high degree by emissions in the brown coal districts in southeast Germany. Circulating winds offered the chance to study thoroughly the short-term interrelations between gaseous pollutants and the physiologically important carbon dioxide which was included in the study at an early stage of the project. In the case of the nitrogen oxides, the influence of the dense motor traffic in the conurbation of Hamburg is obvious. The distribution of ozone exhibits a broad maximum for winds originating from easterly directions. Rather low values observed from the direction of Hamburg may be explained by the partial consumption of this gas for the oxidation of nitrogen monoxide emitted.

The long-term mean concentrations of both sulphur dioxide and nitrogen oxides were found to be below the limits recommended for an optimum protection of the forests. The situation is quite different in the case of ozone. Here, the mean values were above the recommended limits over the entire measuring period. The concentrations of all gaseous pollutants at the "Postturm" site in part clearly exceeded findings from South German forest decline areas. Detailed results are presented on the seasonal and diurnal variations of the gas concentrations and the turbulent fluxes. Particular attention is also paid to the occurrence of nocturnal ozone maxima. Of all the gaseous pollutants measured only sulphur dioxide shows a clear long-term trend with a decrease of about 50% during the investigation period. This trend did not start significantly before 1990 which is in good agreement with the results obtained for the dry deposition of sulphur via airborne particulates. In the case of ozone, an increase in the tropospheric concentration, which has often been reported in the literature, is not apparent. Most probably the evident increase in former decades has come to a standstill as a consequence of the stabilization of the nitrogen oxide and hydrocarbon emissions. The fluxes of the nitrogen oxides do not deviate significantly from zero. A similar conclusion can be drawn in the case of carbon dioxide. The net balance is practically zero as well. This is not in agreement with general experience, since the spruce trees still exhibit distinct growth. On the other hand, there is a pronounced dry deposition of ozone.

Of particular interest are the findings that high concentrations of gaseous pollutants cause marked alterations in the carbon dioxide concentration. The temporal behaviour of the response and the dimensions depend on the pollutant considered and its concentration. While a rise in the concentration of sulphur dioxide, for instance, induces an instantaneous reaction, the results in the case of ozone indicate a correlation between the ozone maximum by day and the carbon dioxide maximum values in the early morning of the next day. These observations suggest that different mechanisms of action may be involved. The fluxes of gaseous pollutants and carbon dioxide have opposite signs, i.e. a marked deposition of pollutants causes a release of carbon dioxide from the ecosystem. It is obvious that a search for the sources responsible begins in the crown compartment as a result of direct effects on the needles. Indeed, several studies have shown that stress caused by gaseous pollutants can induce a significant depression of net photosynthesis and stimulate dark respiration (see, e.g., Küppers and Klumpp 1988; Lange et al. 1989). The present study also suggests that the soil contributes to the release of carbon dioxide with the above temporal behaviour. Quantitative assertions are not yet possible. Likewise, details of the mechanisms involved in the observed total phenomenon must still be clarified. Therefore, it is suggested that further investigations of the direct impact on needles in appropriate fumigation experiments with high time resolution should be performed and that measurements of the carbon dioxide fluxes from the forest soil by automated monitoring (Brumme and Beese 1995) should be carried out under controlled pollution conditions.

The analysis of the carbon dioxide gradients as a function of the gradients of sulphur dioxide and ozone, respectively, reveals that the impact of ozone is much more severe than that of sulphur dioxide. Moreover, the results appear to indicate that the impact of ozone does not depend very strongly on the season. For a number of events, a correlation between the fluxes of sulphur dioxide and carbon dioxide was elaborated. Of particular interest is a computer simulation in which episodes with ozone concentrations and deposition above certain thresholds were eliminated in the long-term mean of the carbon dioxide flux and replaced by mean values obtained under conditions without the influence of ozone pollution. It turns out that with decreasing impact the turbulent fluxes of carbon dioxide improve towards negative values which were observed in the early 1970s in a Bavarian forest stand under favourable environmental conditions. A similar, although minor effect, was also found in the case of sulphur dioxide.

The results of the present study corroborate the hypothesis that the impact of air pollution on forest ecosystems is of a multifactorial nature. Focus of the research work was the investigation of inorganic pollutants. Organic compounds in the atmosphere were also analyzed at the "Postturm" site. The results have been published elsewhere (Dommröse and Figge 1988; Figge and Dommröse 1992). Organic pollutants further enhance the stress effects on the forest ecosystem. A broad spectrum of substances could be identified including acyclic, alicyclic and aromatic hydrocarbons, halogenated hydrocarbons, alcohols, aldehydes, ketones, carboxylic acids and their esters, phenols and heterocyclic compounds.

References

Brumme R, Beese F (1995) Automated monitoring of biological trace gas production and consumption. In: Alef K, Nannipieri P (eds) Methods in applied soil microbiology and biochemistry. Academic Press, London, pp 468–472

Dommröse AM, Figge K (1988) Qualitative und quantitative Bestimmung organischer Schadstoffe in der Luft des Standortes "Postturm", Forstamt Farchau/Ratzeburg. In: Bauch J, Michaelis W (eds) Das Forschungsprogramm Waldschäden am Standort "Postturm", Forstamt Farchau/Ratzeburg. GKSS Forschungszentrum Geesthacht, GKSS 88/E/55, pp 61–79

Dünisch O, Bauch J, Rademacher P, Puls J (1992) Beurteilung der Düngung eines umweltbelasteten Fichtenbestandes im Hinblick auf seine Stabilisierung. In: Michaelis W, Bauch J (eds) Luftverunreinigungen und Waldschäden am Standort "Postturm", Forstamt Farchau/Ratzeburg. GKSS Forschungszentrum Geesthacht, GKSS 92/E/100, pp 251–286

Figge K, Dommröse AM (1992) Organische Spurenstoffe in der Atmosphäre der Waldstandorte "Postturm", Forstamt Farchau/Ratzeburg und "Donaustauf". In: Michaelis W, Bauch J (eds) Luftverunreinigungen und Waldschäden am Standort "Postturm", Forstamt Farchau/Ratzeburg. GKSS Forschungszentrum Geesthacht, GKSS 92/E/100, pp 71–90

Gawel JE, Ahner BA, Friedland AJ, Morel FMM (1996) Role of heavy metals in forest decline indicated by phytochelatin measurements. Nature 381:64–65

Grill E, Winnacker EL, Zenk MH (1988) Occurrence of heavy metal binding phytochelatins in plants growing in a mining refuse area. Experientia 44(6):539–540

Küppers K, Klumpp G (1988) Effects of ozone, sulfur dioxide, and nitrogen dioxide on gas exchange and starch economy in Norway spruce (*Picea abies* [L.] Karsten). GeoJournal 17(2):271–275

Lange OL, Heber U, Schulze ED, Ziegler H (1989) Atmospheric pollutants and plant metabolism. In: Schulze ED, Lange OL, Oren R (eds) Forest decline and air pollution. Ecological Studies 77. Springer, Berlin Heidelberg New York, pp 238–273

Rademacher P, Kriebitzsch WU (1992) Diagnostischer Düngungsversuch an Fichte am Standort "Postturm". In: Michaelis W, Bauch J (eds) Luftverunreinigungen und Waldschäden am Standort "Postturm", Forstamt Farchau/Ratzeburg. GKSS Forschungszentrum Geesthacht, GKSS 92/E/100, pp 287–306

Schat H, Kalff MMA (1992) Are phytochelatins involved in differential metal tolerance or do they merely reflect metal-imposed strain? Plant Physiol 99(4):1475–1480

Subject Index

Printing: Mercedesdruck, Berlin
Binding: Buchbinderei Lüderitz & Bauer, Berlin